VGM Opportunities Series

W9-DDJ-738

WITHDRAWN

OPPORTUNITIES IN
ROBOTICS CAREERS

Jan Bone

Foreword by
Donald A. Vincent
Executive Vice President
Robotics Industries Association

 **VGM Career Horizons**
a division of *NTC Publishing Group*
Lincolnwood, Illinois USA

Cover Photo Credits

Library of Congress Cataloging-in-Publication Data

Bone, Jan.
    Opportunities in robotics careers / Jan Bone.
        p.   cm. — (VGM opportunities series)
    ISBN 0-8442-4057-5 (hardcover) — ISBN 0-8442-4058-3 (soft)
    1. Robotics—Vocational guidance.   I. Title.   II. Series.

 TJ211.25.B66   1993
 629.8' 92' 02373—dc20   92-46256
                                                              CIP

Published by VGM Career Horizons, a division of NTC Publishing Group.
© 1993 by NTC Publishing Group, 4255 West Touhy Avenue,
Lincolnwood (Chicago), Illinois 60646-1975 U.S.A.
All rights reserved. No part of this book may be reproduced, stored
in a retrieval system, or transmitted in any form or by any means,
electronic, mechanical, photocopying, recording or otherwise, without
the prior permission of NTC Publishing Group.
Manufactured in the United States of America.

3 4 5 6 7 8 9 0 VP 9 8 7 6 5 4 3 2 1

ABOUT THE AUTHOR

Like most of you, Jan Bone has been fascinated by the idea of robots ever since she first heard the term. As a technical writer who has toured a number of industrial sites and laboratories, she has had the chance to see robots in action—from the simple pick-and-place robot in a small plastics factory to the giant, full-service workhorses of the automobile assembly lines.

If she were a young person beginning a career, she says, she'd be a manufacturing engineer. Factory automation (and robots are just one component) offers opportunities for those who enjoy high-tech challenges.

Jan's career as a writer, however, spans more than four decades. Ever since her first newspaper job on the Williamsport (Pennsylvania) *Sun,* she has enjoyed the fast-moving world of deadlines and changing topics. Her writing covers a variety of subjects: from celebrity interviews to human interest features in magazines like *Family Circle* and *Woman's World* to technical articles in professional journals like *Safety and Health* and *Hazmat World: The Magazine for Environmental Management.* Her 1989 article and photographs about the Soviet agro-industrial complex were published in U.S. and international editions of *Food Engineering.* She

iii

wrote the "White Paper" for the National Safety Council's Ergonomics Symposium.

She is an associate member of the Society of Manufacturing Engineers (SME) and of two of its divisions: Robotics International (RI), and Computer and Automated Manufacturing (CASA). In addition, she is an associate member of the Institute of Industrial Engineers, and of the American Society for Training and Development.

A prolific free-lance writer, Jan is senior writer for the Chicago *Tribune's* special advertising sections. Her home computers are linked by modem to the *Tribune's* mainframe. Through electronic database searching, she can access files from more than 90 national and international newspapers.

From 1977 to 1985, Jan was an elected member of the board of trustees of William Rainey Harper College in Palatine, Illinois, and served as its secretary from 1979–85. During that time, the college set up a CAD/CAM Center (computer-aided design and computer-aided manufacturing), sparking Jan's interest in factory automation.

Jan is co-author with Ron Johnson of *Understanding the Film* (National Textbook Company, 4th edition, 1990) and author of *Opportunities in Film, Opportunities in Cable Television, Opportunities in Telecommunications, Opportunities in CAD/CAM, Opportunities in Laser Technology Careers,* and *Opportunities in Plastics Careers* in the VGM Career Horizons series.

Since 1983, she has been listed in *Who's Who of American Women*. She has won the Chicago Working Newsman's Scholarship, the Illinois Education Association School Bell Award for Best Comprehensive Coverage of Education by dailies under 250,000 circulation, and the American Political Science Association Award for Distinguished Reporting of Public Affairs.

A graduate of Cornell University, Jan earned her M.B.A. degree from Roosevelt University. She has taught adult education classes in writing for over 20 years and is an instructor in composition at Roosevelt University.

She is married, the mother of four married sons, and grandmother of Emily Diane and Jennifer Marie.

ACKNOWLEDGMENTS

The following individuals were especially helpful in the development of this book: Hadi Akeel, Louise Bacon, Gilbert Bandry, Valerie Bolhouse, Allan Chrenka, Nick Ihnat, Jim Lakatos, John Miesen, Ron Potter, Paul Schenker, Susan Steelman, Donald A. Vincent, Shelly Weide.

The author also wishes to thank the following organizations for their assistance: Accreditation Board for Engineering and Technology, American Association for Artificial Intelligence, American Association of University Women, Automated Imaging Association, Committee on the Status of Women, American Physical Society, Global Automation Information Network, Institute of Industrial Engineers, International Federation of Robotics, Jet Propulsion Laboratory, Junior Engineering Technical Society, National Occupational Competency Testing Institute, National Service Robot Association, Robotics Industries Association of America, Society of Manufacturing Engineers, Society of Women Engineers, SPIE—The International Society for Optical Engineering, University of Michigan Transportation Research Institute.

FOREWORD

American companies, whether in manufacturing or services, face greater worldwide challenges than ever before. The importance of improving quality, productivity, and cost efficiency continues to grow, and we can expect a greater reliance on robotics and automation in every industry. Robotics is an essential ingredient in strengthening the manufacturing capability of the United States.

We have a very long way to go before robots are used in the kind of numbers that they should be in North America. Successful utilization of any advanced technology, especially robotics, requires a long-term commitment. Foreign companies have shown their willingness to invest in robot applications and advanced automation. Manufacturing managers in our country face constant global competition and sense the need for change. As management attitudes change and more successful robot applications emerge, we will see a steady increase in the number of robots installed in American factories.

Investigating career opportunities in robotics requires a realistic attitude. Books like this one provide the realism, direction, and outlook for considering a robotics career. Most jobs will be found in companies using robots in industries such as automotives, electronics, aerospace, food processing, pharmaceuticals, chemicals, textiles, furniture, and appliances.

Employment in microsensors, microrobotics, and artificial intelligence is on the horizon. The challenge is to develop engineering talent and skills to work in these specialized fields, and then aggressively seek a career. The robotics industry is on the right track and poised for growth. With growth comes new careers, and a career in robotics can be a rewarding one.

> Donald A. Vincent
> Executive Vice President
> Robotic Industries Association

CONTENTS

CHAPTER 1

ROBOTICS:
AN IMPORTANT TECHNOLOGY

From the lovable R2D2 of *Star Wars* to the massive overhead gantry robots of an assembly line, robots have fascinated and intrigued people. The vision of an automated factory in which robots and other machines work together to turn out products at high speeds has moved far closer to reality than Czechoslovakian author Karel Capek believed possible when he coined the word *robot* in his 1921 play, *R.U.R.* Capek took the term from the Czech *robota*—a word that meant a serf, or one who did subservient labor.

TWO KINDS OF ROBOTS

Industrial Robots

In today's industrial world, robots do far more than work on assembly lines. Robots with grippers perform tasks in such fields as die casting, loading presses, forging and heat treating, and plastic molding. They load and unload other machines. A different kind of robot—one that can handle a tool instead of grippers or uses its grippers to grasp a special tool—is used in applications

1

like paint spraying; spot or arc welding; and grinding, drilling, and riveting in machining.

At Lockheed's giant Sunnyvale, California, plant, robots are used in assembling printed circuit boards. In Japan, robots put tiny screws in watches, tightening them automatically in place. From aerospace plants to automotive assembly lines to appliance manufacturing, robots are helping industry increase production rates while keeping quality control constant.

Robotic Industries Association (RIA), the North American trade group that focuses exclusively on the robotics industry, estimated that in 1992, some 44,000 robots were in use in the United States. Roughly half were being used in the automotive industry. Other leading user industries included appliances, electronics, food and beverage manufacturing, pharmaceuticals, and heavy equipment.

Nonindustrial Robots

Yet there is another side to robotics technology—a side you may be surprised to learn about. Robots are used in nonindustrial settings. "Wherever people want their skills augmented, such as in helping the developmentally disabled, performing security functions, or relieving humans from hazardous jobs, robots are beginning to appear on the scene," says the National Service Robot Association.

Service robots have many applications. For instance, "Robodoc," a robot representing a unique surgical application of robotics, has performed 25 successful hip replacements in dogs. The California-based manufacturer has asked the U.S. Food and Drug Association to approve the device for human hip replacement surgery.

In Japan, a university has developed a robot that simulates the jawbones of humans. Researchers hope to use the robot to help them understand how the human chewing mechanism works. Then they can devise appropriate dental treatment.

Out of the research phase and already in commercial production in Cambridge, England, the Inventaid Wheelchair Manipulator is helping quadriplegics (persons who have lost the use of their arms and legs). The Manipulator can lift a briefcase, open a door, pour a drink, and lift the drink cup to the mouth of the disabled person.

ADVANTAGES OF ROBOTS

Nonindustrial robots can work in environments where humans can't—or don't want to. For instance, NASA scientists plan to send an eight-legged robot, developed by Carnegie-Mellon University, 700 feet into an active Antarctic volcano about 800 miles from the South Pole. The project, funded by NASA's telerobotics program, has two goals: to test the robot for future missions to Mars and to get a closer look at minerals and gases within the volcano. The robot's sensors include a laser scanner with a 350-degree field of view, giving on-board computers information about where to put the robot's "feet." When it reaches the base of the crater, the robot is programmed to suck up samples from gas vents through a hollow probe.

In Connecticut, a robot manufacturer is developing a prototype robot to clean washrooms. Funded by a joint grant from the state of Connecticut and the U.S. Post Office, the project aims at making the robot an assistant custodian. The robot will be programmed to clean and sanitize toilets, sinks, partitions, walls, and floors.

Service robots developed by Cybermotion are used to patrol buildings. They each travel over 10 miles a night, allowing security officers to be free for additional duties.

Service robots and related devices have saved companies substantial sums of money, according to the Electric Power Research Institute (EPRI). For instance, a remotely operated underwater vehicle helped Public Service Electric and Gas Company with problems encountered during a refueling outage at a nuclear station. The device helped move the internal components of the reactor vessels. EPRI reported the company saved about $635,000 the first time the device was used.

Another EPRI report has shown that robots can be used for 29 potential tasks in handling solid radioactive waste, decontaminating nuclear reactors, and handling filters. Robotics and related technologies can be used in remote devices to make workers safer by limiting human radiation exposure.

BACKGROUND OF ROBOTS

Questions about manufacturing—and whether workers would lose jobs through improvements in technology—have probably been around since 1801, when French inventor Joseph Jacquard worked out a textile machine operated by cards with holes punched into them. The machine could quickly and easily produce the intricate designs that soon became the height of fashion. Just 12 years later, 11,000 automated looms were operating in France.

Other names and dates stand out in the history of automation: Christopher Spencer's cam-operated lathe in 1830, Seward Babbitt's motorized crane with grippers that removed ingots from a furnace in 1892, and Willard Pollard's and Harold Roselund's programmable paint-spraying mechanism in 1938.

Shortly after World War II, George Devol patented a general-purpose playback device for controlling machines—a device that used a magnetic process recorder. Also in 1946, the first large computer—the Eniac—was built at the University of Pennsylvania, and another large computer—the first general-purpose digital computer—was introduced at Massachusetts Institute of Technology (M.I.T.).

Eight years later, Devol designed the first programmable robot, naming his technology Universal Automation. The name eventually was shortened to Unimation, which became the name of the first robot company.

The industry grew slowly. Veteran engineer Roy Morley, who spent much of his career with Caterpillar Tractor Company in Peoria, Illinois, remembers how in 1975, as a brand new engineer in the company's orientation program, he helped to install Caterpillar's first robot—"the 15th industrial robot shipped by that manufacturer, I think."

General Motors installed its first industrial robot on one of its production lines in 1962. The company's move to robots in any quantity started around 1970. But even as late as 1980, GM had only about 275 robots, so the first years were very slow.

The year 1980 was the time when robots started to take off, when the industry began to grow rapidly, when optimism rode high, and a career in industrial robotics sounded like a chance to get in on the ground floor of factory automation.

In the early 1980s, growth rates in the industry were impressive. A U.S. Department of Commerce report on "A Competitive Assessment of the U.S. Flexible Manufacturing Systems Industry," prepared in 1985, summarizes those years. U.S.-based robot vendor sales totaled $90 million in 1980, $155 million in 1981, and an estimated $190 million in 1982. In 1984, Prudential-Bache Securities estimated 1982 sales at $375 million and 1983 sales at $240 million.

In the mid-eighties, however, robotics companies hit a downturn in sales. As *Managing Automation,* one of the trade publications that includes news about robotics, puts it, "Reality set in, as the technologies' limitations clashed with manufacturers' expectations." Don Arney, division chair, New Technologies, at Indiana Vocational Technical College, agrees that robotics may have been oversold in those years.

"When we first began to look at the robotics field in the early eighties," Arney says, "we felt the growth rate of the robotics industry had been overestimated by a considerable margin. As a result of industry surveys, we discovered that a far more important area of growth was in the general field of automation, where robots were merely another 'automated tool' to be used within the manufacturing endeavor." Based on these surveys, Indiana Vocational Technical College set up its Automated Manufacturing Technology Program—a program which placed 100 percent of its 1991 graduates at an average starting salary of $19,760.

Perhaps the trend is turning, however. In 1991, the Robotic Industries Association (RIA) reported that U.S.-based robotics firms shipped nearly 4,500 robots—the most since 1986.

RIA's executive vice-president, Donald A. Vincent, credited the increase to the fact that a wider range of customers had begun to benefit from robotics. In the first half of 1992, RIA statistics showed 2,600 new robots ordered—the highest first-half total since 1986. Vincent also predicted factories would begin to increase their investments in production automation technologies as the United States and other countries began to recover from economic problems.

Vincent cautions, however, that the U.S. robotics industry has "a very long way to go" before robots are sold in the kind of numbers that they should be in North America. "Only in Japan, where some 50,000 robots are installed *each year,* are companies taking full advantage of the productive power of robotics," he says. "It's no accident that Japan is such a strong competitor in so many manufacturing industries; their companies are willing to make a long-term commitment to robotics and advanced automation."

TWO FACTORS TO KEEP IN MIND

If you are considering a career in robotics, there are at least two important things for you to think about.

One is the difference between "estimated" and "actual." Analysts can take past growth history and project it to estimate what they believe sales and shipments will be in a future period. Whether they are right, however, depends on many factors outside their control.

Overcapacity is one of those factors. General Motors Corporation has announced it will reduce its work force at its North American operations by 74,000 employees. Target date for the new, lean look? "Mid-nineties," according to GM spokesman Mark Tanner. The rise of the global work force, as *Fortune* terms the world's supply of skilled labor, will change the locations at which companies build new factories and invest in automated technology. The magazine predicts that "more than ever before, work will flow to the places best equipped to perform it most economically and efficiently."

The second thing you need to keep in mind, particularly when you read worldwide comparative figures, is that not everyone defines *robots* in the same way.

Many countries and users have accepted the definition of the Robotic Industries Association (RIA), a U.S.-based trade association of companies that use or are considering using robotic equipment as well as companies that manufacture or market robotic equipment. RIA defines *robot* as "a reprogrammable, multifunctional manipulator designed to move materials, parts, tools or other specialized devices, through variable programmed motions for the performance of a variety of tasks." In chapter 2 of this book, we will look at just what that definition means.

The Japanese define *robot* differently. Since 1979, the Japanese Industrial Standards have classified industrial robots according to the method by which information is given to the robot and how the robot is taught. In fact, as much as a third of what the Japanese call robots may not be robots at all by the RIA definition.

IMPLICATIONS FOR YOU

What does this information mean to you if you are considering robotics as a career?

First, it means that there are currently fewer robots in U.S. factories than you probably realized. In fall 1992, RIA estimated there were approximately 44,000 robots, all told, in the United States.

Second, it means that a degree or certificate in robotic technology is probably not automatically going to guarantee you a job in robotics. (See later chapters in this book for more about the issues of what you need to know in order to work in the industry and whether the supply of graduates is outstripping the jobs available for them.)

Finally, it means that if you want to work in robotics, you will have to take a good deal of responsibility for learning what is going on in the industry and related fields. You yourself will have to keep up with developments. Reading professional journals, joining student chapters of professional societies and taking an active role in membership, and staying on top of trends become crucial. If you are serious about robotics—and related fields—you cannot afford to ignore reality. In order to make appropriate choices, you will have to keep up with business and economic developments and modify your plans accordingly.

SCOPE OF THIS BOOK

Because robotics must be considered as just a part of program-mable automation and because of the interplay of economic factors, this book cannot cover the field comprehensively. Instead, materials about schools, associations, and periodicals—along with information about robotic technology and stories from people currently working in robotics—will help you learn where to find information.

Do you want a job in robotics? Can you get one? These are the questions that *Opportunities in Robotics Careers* will help you answer. There are no easy answers in today's job market, but this book will give you information about the industry's various components, places to go for more details, and an awareness of how a few people in the industry view their jobs.

CHAPTER 2

HOW A ROBOT WORKS

When we look closely at the definition of a robot, we begin to get some idea of how this machine works and of its amazing versatility. The Robotic Industries Association (RIA), a trade association of robot manufacturers, consultants, and users, defines an *industrial robot* as "a reprogrammable multifunctional manipulator designed to move material, parts, tools or specialized devices through variable programmed motions for the performance of a variety of tasks."

A robot, then, can be programmed—and programmed again and again. It is multifunctional, able to do more than one or two tasks. The robot is a manipulator, with an "arm" that moves mechanically, grasping and moving material, parts, tools, or devices in a series of predetermined motions to achieve the desired objectives. These may vary by industry, by job, or by specific task.

Three basic parts make up a robot: a manipulator, a power supply, and a system for controlling the robot.

THE MANIPULATOR

The robot arm, from the base of the robot through the wrist, is called the *manipulator*. Within that arm, there are various components such as actuators, drives, bearings, and feedback. These make it possible for the arm to move in different directions.

The extent to which the robot hand, or working tool, can reach in all directions is called its *work envelope*. The dimensions of the work envelope depend on the way the manufacturer has put together the robot's axes of motion, or degrees of freedom.

Usually, a robot has three major axes of motion: a *vertical axis,* which determines how high the robot can reach; a *horizontal axis,* which lets the robot move in and out; and a *swing* or rotation around the robot's base.

Some robots have additional axes of motion. These often are referred to as *pitch, yaw,* and *roll.* A "wrist" at the end of a robot's arm contains components that allow the additional movement.

Many robots currently available are designed in a modular fashion. That design lets users choose the number of axes they want. They usually make that decision based on how they plan to use the robot and the amount of payload they need. (*Payload* is a term that means the robot's ability to carry a given maximum weight at a given speed continuously. It is expressed in units of weight, such as pounds or kilograms.) Manufacturers who want to increase a robot's reach and ability to manipulate objects can provide additional axes of motion in end-of-arm tooling.

THE POWER SUPPLY

Three different types of power are used for industrial robots: pneumatic, hydraulic, or electric. Robots that have a light payload are often powered pneumatically. Robots that need to carry very heavy payloads often use hydraulic power. However, electrically powered robots are becoming more popular. Electric robots have several advantages over hydraulic robots: they do not leak, they are usually quieter to operate, and they usually use less energy. An electrically powered robot is "ready to go"—that is, it does not need to be "exercised" before production begins, as hydraulically powered robots usually are, to get the fluids up to normal operating temperature. Engineering research is helping manufacturers downsize electric motors so electrically powered robots can carry larger payloads. Then electric robots will be able to have larger payloads.

THE CONTROL SYSTEM

Non-Servo Point-to-Point Robots

Pick-and-place robots, as Modern Machine Shop's *Machine Tool Handbook* describes them, are perhaps the simplest class of robots. A robot like this has a pneumatic or hydraulic cylinder or motor (an actuator) connected to a valve that controls the robot's direction. An input signal to the robot shifts the control valve. Then, fluid flows into one port of an actuator for an axis of motion. The other port is vented; the fluid is returned to the supply tank. The worker adjusts the stroke of each axis of motion by adjusting mechanical stops to the desired positions. A sensor, mounted on each axis of motion, can be adjusted to determine the two end-points of each axis stroke.

Such a robot is usually programmed mechanically for a particular application. The robot operator determines what motions the robot should make and in what sequence. Then the mechanical end-stops for each axis are adjusted to the appropriate position.

Servo-Controlled Point-to-Point Robots

Servomechanisms (closed loop) can stop a robot along each axis of motion at an indefinite number of points. As the axes approach a preprogrammed point, the control system slows down the robot. Eventually, the control system closes the servomechanism and stops the robot at the desired point.

Programming of such a robot is usually done with a *teach pendant*—a control box that the operator holds, similar in shape and appearance to the remote control of a television set. Points in space are stored in the robot's memory. Then the robot carries out the programmed sequence of motions by proceeding from point to point.

A robot like this is often used in tasks involving loading and unloading.

Servo-Controlled Continuous Path Robots

Both servo-controlled point-to-point robots and servo-controlled continuous path robots use feedback devices on each axis of motion. *Continuous path robots* (often used for spray painting and other finishing operations) are taught by an operator who actually leads the robot arm through a pattern of motions. Meanwhile, the robot's control system uses various feedback devices to tell it what position the various axes are in. The data from the robot's sensors are recorded and stored.

A continuous path robot often requires much more computer memory than a robot that uses point-to-point programming.

PROGRAMMING THE ROBOT

In order to do its work properly, a robot must be programmed. There are several ways in which this can be done.

Teaching by Guiding

Some robots are programmed by having an operator guide them through the desired sequence of tasks. Typically, a servo-controlled point-to-point robot is programmed using a teach pendant.

Each one of the robot's axes is servo-driven (a closed loop system). The robot has the ability to stop and start at any point.

Attached to the robot's axes is a feedback device—like an encoder—that gives feedback on the position of an axis to the control.

On the teach pendant, there is a button for each axis to move up, down, left, right, or rotate. The operator starts with the robot back at its home position, ready to go. Pushing the button puts the root in the "teach mode."

Suppose we want the robot to move out, grip a part, lift the part out of a machine, retract its arm, rotate its arm over, move down, ungrip the part to set it on the fixture, retract the arm, and rotate it back for another cycle.

The person teaching the robot pushes the button for the "reach" motion and watches the robot. Holding down the button moves the robot arm. When the arm gets to where the operator wants it to be, he or she releases the button.

Next, the operator pushes the "record" button. Using all the feedback devices, that function assesses the position of each of the axes and puts those positions in memory as step 1.

Now the operator advances the control to the next step. If the robot is supposed to grip the part, the operator pushes the teach pendant button that says "grip." The robot grips the part. Then the operator presses the "record" button. From here on, the robot knows it has to grip at step 2 in the procedure.

It's the same procedure for the other motions. The operator walks the robot through the program steps slowly, making sure to record each one. When the operator is finished, he or she can switch the robot from "teach" to "repeat."

The comparator in the robot controls compares where the robot is going to where it actually is as it is approaching the point to which it has been programmed to go.

Another way of teaching by guiding is used for servo-controlled continuous path robots. Many of these robots are used in spray painting and finishing tasks, where typically a smooth, continuous motion is needed. In order to teach such a robot, the operator physically takes the end of the robot arm and leads it through a pattern of motions. Meanwhile, the robot's control system is recording feedback data from the position sensors on the axes and storing them on a mass memory storage system.

Unlike the point-to-point robot, the continuous path robot stores its data on a "time" basis instead of storing it as a series of individual discrete points. However, both types of robots use similar feedback devices on each axis of motion.

Teaching by guiding has limitations. It is not always easy to edit a path—that is, to change only one portion of the path without having to rerecord the entire path.

As an analogy, consider a secretary who has typed a letter on a personal computer, using word processing software. If the letter has to be changed, it is simple to retrieve that particular file (stored on a floppy disk or the hard disk drive of the computer), add or delete a few sentences, close the file again with a computer command, save the edited version, and print the revised letter.

It is easy and quick for the secretary to produce the edited version of the letter, because the word processing software allowed changes to be made. The entire letter did not have to be rekeyed into the computer.

Some teach-by-guiding robots have programs that can be edited easily. Others don't.

Off-line Programming

Another method of programming robots is called *off-line programming*. Usually, a computer operator writes a program while sitting at a terminal—not necessarily on the factory floor. The program is given to the robot to follow, and the robot carries out the instructions the operator has written.

One robot expert describes off-line programming this way: "Our programming runs on an IBM-compatible PC or on a terminal to a DEC MicroVax or large VAX. Off-line programming is editing a program that tells the robot where to move and what to do when it gets to that place. You tell the robot how fast to go and its orientation with respect to the tooling the robot carries.

"You use a standard computer editor and write a program in English with XYZ coordinates of where to go.

"After the commands that tell the robot to do a program have been edited, the program is run through an off-line programming module that looks for errors. The module then reformats the program—that is, puts it in a form that the robot will understand. The program can be sent to the robot over a communications line, or it can be put on a floppy disk that is taken to the robot and loaded into the robot controls.

"Off-line programming has several advantages. An important benefit is that the assembly line does not need to be shut down while the robot is being programmed. Another plus is that the operator can create a program that instructs the robot to do complex tasks. Robots are often called to sense their external environment and to make subsequent decisions, such as, "If the part you are to pick up is not where it should be, wait for the next part."

Once the robot has been taught a series of operations, the data are stored on a flexible disk as a program that can be changed later. The program's editing capabilities allow the operator to improve the robot's recorded motion by cutting out "overspray" (if the robot is painting) and "deadtime." Three editing functions are available (in this robot) to modify and perfect programs: end-to-end assembly, deletion, and insertion.

Depending on the end user's needs, a manufacturer may offer robots that can be taught by guiding or programmed off-line. Some manufacturers are able to combine off-line programming with computer-generated data. Still other consultants and experts who study the robot industry believe changes are coming in the ways in which robots are programmed and controlled.

As reported in *Managing Automation,* a leading trade publication that covers computer-integrated manufacturing (CIM), companies are beginning to simplify software and controls. Some manufacturers have already developed automation controls that "understand" their customers' manufacturing processes—from part feeding and pallet geometry to conveyor tracking and work-cell control. They have coupled these controls with software that gives users quick, easy access: simple, pull-down menus, step-by-step prompts, and point-and-click commands. Some process adaptive controls can be retrofitted to existing multi-axis robots, thus saving time and money; one manufacturer reports savings of half a million dollars on a single automation project.

Another trend that is becoming increasingly important is for companies to offer robot accessories: grippers, pneumatic slides, and conveyors. As Nick Ihnat of Jergens, Inc., puts it, "Automation components for handling and assembly make sense. If a manufacturer needs to grab, lift, turn, transfer, rotate, convey, or reposition parts, it's easy to add grippers, slide assemblies, or pick-and-place manipulators. Companies like ours deal primarily with the automotive and electronic industries—industries in which there's mass production and manufacturers can set up dedicated equipment to turn out product."

Fully automated inspection systems that reject defective parts are also becoming popular, according to *Managing Automation.*

Machine vision, the magazine says, is becoming more cost-effective, easier to use, and easier to maintain.

For both robotics and machine vision, says magazine Editor in Chief Robert Malone, "The technology is catching up, new applications are developing, and the industry looks forward to better days."

ROBOTS WORK WITH OTHER MACHINES

One of the most important things to remember about robots is that they are just one part of technology used in automation. Robots—whether they are industrial robots or service robots—need to be looked at as part of a larger system rather than as stand-alone machines. "I would not recommend that a student pursue a career just in robotics," warns Dr. J. T. Black, from Auburn University's Advanced Manufacturing Technology Center. "It's too narrow. Also, increasing numbers of robots are being designed and manufactured in Japan. Instead, the jobs are in learning how to apply robotics in manufacturing environments."

In other chapters of this book, we will talk about larger systems and what the implications of Black's warning are for job prospects.

CHAPTER 3

MOVING TOWARD AUTOMATION

If you are like most people, you have had very little chance to see what goes on in a manufacturing plant. Perhaps you have taken a tour of such a plant with your family or with one of your school classes. Possibly one of your family members or neighbors works in manufacturing. But many people have never had an opportunity to see firsthand the challenge and opportunity a career in manufacturing offers.

Part of the reason for this lack of knowledge is that insurance rules and fear of legal liability have led many industries to bar young people. In most manufacturing operations, persons under 16 or 18 years of age (the rules may vary from factory to factory) cannot be on the plant floor, at least while machines are running. It is a situation that concerns many top executives.

MODERNIZING MANUFACTURING

"If the United States is going to maintain its lead as one of the world's great producers, we must interest young people in making a career in manufacturing," declares Robert L. Vaughn, a former president of the Society of Manufacturing Engineers. "Otherwise, I think our country is going to come to a halt."

Vaughn says today's manufacturing is "not like beating with a hammer on a piece of metal.

"Yes, you can still do that," he explains, "and some people will always want to, but modern manufacturing is all computers."

According to *Fortune,* in the two-year period of 1991 and 1992, businesses boosted outlays for computer power by 70 percent. At the same time, manufacturing productivity jumped 4.2 percent, despite a sluggish economy. Software programs, once affordable only by large companies using mainframes, are becoming available at modest cost to run on personal computers.

Manufacturers, too, are looking at modernizing their equipment. Harry Matthews, director of operations management at A. D. Little, a well-known consulting firm, says production advances in the eighties came from better engineering and improved worker effectiveness. He warns, however, that production technologies in place at many manufacturers will not keep them competitive to the year 2000. As a result, small to mid-sized companies are looking seriously at automation. That may mean more robots—and more opportunities for those who would like to work with them.

COMPUTER-CONTROLLED AUTOMATION

What Vaughn is referring to—an extremely important concept for you to understand if you hope to work in robotics and related technologies—is often called "programmable automation." It is a term that refers to using technology that can be programmed by a computer.

Robots and robotic technologies are part of *programmable automation,* a way for improving manufacturing that many people believe will help U.S. manufacturers to become more productive and globally competitive. The idea of using technology to streamline manufacturing processes is not new. In fact, the first wave of automation technology took place in the 1950s and 1960s. Now, just as then, a number of questions and concerns arise—concerns of interest to you as you explore the idea of working in robotics.

- Will the new technologies put a significant number of people out of work?
- Will the introduction of new technologies dehumanize the work environment?
- How can the United States best prepare its educational and training system to respond to the growing use of computerized manufacturing automation?

A GLOBAL LABOR FORCE

As you look at career options, a new trend you need to consider is the shift by a number of U.S. companies to manufacturing abroad. "Some of the most Japanese-looking American plants are going up in Brazil," says Martin Anderson, a vice-president specializing in global manufacturing for Gemini Consulting, a New Jersey-based firm. 3M makes tapes, chemicals, and electrical parts in Bangalore, India. Hewlett-Packard assembles computers in Guadalajara, Mexico. General Electric designs advanced lighting fixtures in Budapest, and its Hungarian factory is GE's leading center for making advanced compact fluorescent bulbs. AT&T makes telephones in an industrial park on Batam Island, an Indonesian island across the Strait of Malacca from Singapore. And in Singapore itself, Motorola designs and manufactures pagers—one of which accepts voice messages.

It isn't only American companies that are moving to new production sites. ABB, a Swiss-Swedish builder of systems that generate electricity, expects to have more than 7,000 workers in Thailand by the year 2000. Siemens's U.S. subsidiary uses Indian-owned programmers at Indian-owned software companies.

No one is quite sure what this shift to a global labor force means to jobs, and how fast—or even if—it will happen. However, it is certainly likely that factories abroad may be built with the latest in technology and automated equipment. Robotics and related technologies, then, may prove to be increasingly important in future manufacturing.

No easy answers exist for these questions; no consultant or economist can accurately and definitively forecast what will happen. You won't find a well-laid-out blueprint toward a successful robotics career path in this book or elsewhere. But you will be better prepared to examine your opportunities for working in robotics if you are aware of these concepts and trends. You will want to keep in mind just how robots fit into programmable automation. You will want to monitor what manufacturers are doing about outsourcing (the term used when companies contract for parts and labor from independent local suppliers). In short, you will need to take a proactive role in managing your own career development.

CHAPTER 4

PERSONAL QUALITIES

What personal qualities do you need to succeed in robotics and related careers? What do people who are working in these fields believe are the advantages and disadvantages?

WILLINGNESS TO BE TRAINED

Many people in the profession believe that in technical fields, knowledge is advancing so rapidly that what you have been taught in college or vocational schools will be superseded quickly. In fact, some of them say, even if you hold a four-year engineering degree, approximately half of what you were taught will be out of date in seven to ten years. If they are right, you will have to work very hard at keeping up with technology. You will need to be the kind of person who recognizes the need for life-long learning and is willing to spend the time and energy necessary to bring yourself up to speed. That may mean enrolling in continuing education courses or attending workshops and seminars. It almost certainly means reading the trade publications and technical books. You will need not only to learn the basics but also to recognize that those basics will be changing rapidly. You will need not only the ability to be trainable—but also the desire to keep learning.

GOOD COMMUNICATION SKILLS

In chapter 8, you will learn more about what a blue ribbon task force believes are the necessary skills. The information from a 1991 U.S. Secretary of Labor's report on achieving necessary skills represents a move toward competency-based learning—learning that tests what you can actually do rather than assuming you are proficient because you have studied certain subjects in school. One essential quality you will need to succeed in robotics and related technology is good communication skills.

Ford Motor Company's Valerie Bolhouse, author of *Fundamentals of Machine Vision* and a leader in robotics-related professional associations, puts it this way: "You have to be able not only to do programming, but to explain it to individuals who have varying levels of technical skills. You have to be able to communicate effectively what you have done, and what the results were, in a manner that anyone can understand. For instance, it's hard to explain to someone who doesn't have your technical expertise. Yet it has to be done. When two people speak the same technical language, it's easy to communicate—but if the person you're speaking to isn't computer-literate, explaining to them what you can do with a computer and why you can't do their project becomes substantially harder. It's important to use technical language fluently."

Almost certainly, if your job involves robotics, you will have to write reports about what you are doing. Sometimes your skill at communicating may be an important factor in getting approval for a desired project. Increasingly, companies are requiring feasibility studies (Will a project be practical? Will it work? Will the payback period justify making the capital investment required?) before agreeing to spend money on equipment. Your report, explaining that a project is worthwhile or describing the benefits of the work you and your department have accomplished, may be read by people who have no previous ideas about the topic—people who may be in a position to approve or deny expenditures. Your communication skills and your ability to write persuasively may make the difference.

CREATIVE LISTENING SKILLS

All robotics-related jobs aren't technical. Someone has to be out there listening, learning what customers want and need, helping to solve customer problems. Like Nick Ihnat of Jergens Inc., who says he is thought of as "the automation specialist" inside his company (which offers grippers and other end-effectors for industrial robots), you need to be extremely sensitive to customers and even to people in your own organization. You need to hear—creatively—what they are saying. Is your company doing what customers want and need? Are you sure?

Ihnat takes creative listening a step further. He reads his competitors' catalogs cover to cover. Then he asks his customers how Jergens's products measure up against the competition. He uses the answers to help his company keep ahead.

PATIENCE

Another quality Valerie Bolhouse recommends, especially for research and development work in robotics, is patience. "You have to make sure you don't give up too soon, and tell a manager you can't do a project," she points out. "You need to be sure that no matter what it takes in time and effort, you don't overlook an opportunity to save the company money and improve the quality of the product. In addition, you need to be sure that when you devise a solution, it will work over a broad range of products, rather than just being a lab-based solution."

GETTING ALONG WITH ALL LEVELS OF PEOPLE

One quality that is important for success in robotics and related technology is the ability to interface with all levels of people. As Gilbert Bandry puts it, "You may have to interface—not only with top-level executives, but with the plant maintenance personnel—the people who tighten nuts and bolts and get things to work.

"If you are natural, if you are yourself at all times, you can talk to everyone on a human level. You don't have to distance yourself from anyone. You have to be able to grasp concepts quickly, and express your ideas in a very clear, concise manner so that no matter whom you're talking to, no matter what their level of understanding is, you can get your point across."

You may need these skills—not only in English but also in other languages. The United States had roughly 44,000 industrial robots in early 1993, according to Donald Vincent, executive vice-president of the Robotics Industries Association. Yet Japan is installing 50,000 industrial robots a year, he says. You may need to learn Japanese—or other languages—in order to keep up with technical developments or to work with other professionals. Indeed, as we move increasingly toward a global economy, you may even have a robotics-related job in a country other than your native one.

LOOKING AT EXISTING TECHNOLOGY IN NEW WAYS

Creativity—the ability to understand what is happening in robotics and related fields and not to be limited by apparent barriers—is essential, many industry veterans say. "Throughout history, says Ron Potter, a recognized industry leader, there have been mechanical tinkerers who like to deal with machines or mechanisms. They have a peculiar kind of curiosity about how things work, or about how they can make things work—better."

Potter believes the industry can use people with a vision of what they want to do. The industry is changing so rapidly that by the time you finish your technical training, you may find job opportunities far different from what you imagine right now.

Whether you work in industrial robotics or become involved with service robots, whether you specialize in machine vision or artificial intelligence, you will want to look at the total picture—and recognize that it is going to change. If you can keep your enthusiasm and interest high, you may be like those industry veterans who have stayed in robotics during its ups and downs. Virtually all of them are still excited about robotics and love it. "Robotics still is a big turn-on for me," confesses Potter. "Robotics (and other aspects of automation) touch every area of industry involved in manufacturing. Those of us in it still have the desire to excel."

CHAPTER 5

ROBOTS IN INDUSTRY

The totally automated factory of the future does not exist—yet. But companies are moving more and more toward automation. "There's a 'lights out' soap factory down the road from me," says Jim Lakatos, employment counselor in Grand Rapids, Michigan. "It runs 24 hours a day, but usually no one is in there. Robots and automated machinery are doing it all.

"Railroad cars with materials arrive at one end of the factory. Detergent and additives are automatically unloaded into big bins. They feed through conveyor systems into mixing operations. Once the soap is mixed, it moves automatically to filling machines. Product is pushed into boxes, and bar code is put on the sides. Automatic stacking/retrieving equipment loads the boxes on trucks that show up at the other end of the factory and depart with product, heading to stores. Although people come in from time to time to make sure the machinery is running smoothly, no one has to be there at any given point. The factory just goes on making products."

Most U.S. factories have not reached this point, and it is doubtful that they ever will. However, most large industrial corporations have moved into flexible automation. They have invested in technology and retrained personnel in order to stay competitive in the global marketplace. Many of them have reduced labor costs substantially—some, to as low as ten percent of the total cost of production.

Companies such as Motorola, Inc., have refined a number of their operations to take advantage of automation technology. At Motorola's Boynton Beach, Florida, pager plant, robotic "hands" assemble tiny pager components too small for human figures. Sophisticated software coordinates the operation.

At the company's world headquarters in Schaumburg, Illinois, a tabletop automated factory simulates operations at the Florida pager plant. In an adjacent room, machines "reproduce" a portion of the Florida plant. There, Motorola employees learn how to program robots, solve automation equipment problems, and use advanced technology to design better products.

COST FACTORS

However, even now, in the early nineties, over 95 percent of industrial firms in the United States have not automated. Virtually all of them are small- to medium-sized companies. Regulations such as those from OSHA and EPA, the high cost of U.S. labor (and benefits, including health care), and other factors make it difficult for them to maintain the profit margin to which they have been accustomed.

Few of them can make the investment—or even think it a desirable choice—to retrain employees to take advantage of state-of-the-art automation technology. That unwillingness to change has brought some harsh words from Massaki Morita, president of Sony U.S.A.

"When final assembly was the primary focus, as it was some 20 years ago, low wages and rapid turnover were accepted as necessary aspects of manufacturing," Morita told a 1991 conference of American and Japanese CEOs. "The growing importance of precision component manufacturing and assembly—often with extremely sophisticated factory automation technology—brings with it the need for a stable, well-trained and flexible workforce, capable of operating robots and other precision equipment efficiently."

Chicago *Tribune* technology writer Jon Van, who won the twentieth annual Science-in-Society Journalism Award from the National Association of Science Writers, summed up manufacturing problems in his 1991 series, "Recrafting America."

Says Van, "Today's successful manufacturer needs employees who can perform such multiple tasks as operating robots, identifying symptoms of defective robots, and performing simple maintenance. Most of all, workers must understand the production process and work with robots as a team. They must become part of the production process."

Automation (including robotics and related technology) represents an opportunity for small to medium American companies to modernize plants and equipment. But one of the drawbacks toward spending is the way in which payback (the number of years the company requires to recover its original investment from net cash flows) is figured. Although the payback period is easy to calculate, using the payback method to decide if an investment is worthwhile can lead to some wrong decisions. Many expensive projects—including the installation of a robot—can generate income beyond the payback period. Automation always involves a capital investment that must be amortized through cost savings on each unit produced. Because investing in a robot may not yield a return for a number of years, a small business may be reluctant to spend the money required or unable to convince a lender to finance the investment.

Increasingly, robots are being used by smaller factories. Often, they find that labor savings and dependability can offset their initial investment in automation. One such plant, located in Mc-Henry, Illinois, about 50 miles northwest of Chicago, is Alpha Plastics—a job shop deliberately designed to change, retool, and adapt machines and workers to produce short runs of many individual products. The company's custom-produced, heat-sealed vinyl items are sold across the United States and, to a limited extent, in European markets.

Some of the approximately 30 employees on the factory floor operate machines that require them to move plastic pieces back and forth by hand. Others work on machines with turntables that automatically rotate to bring fresh materials or take product away. One such machine, with two workers, has a nearby robot that is programmed to pick up the material and lay it at the same spot every time. "The robot's efficient," says supervisor Marge Kuhms, "and it eliminates the need for one person."

John Miesen, who owns Alpha Plastics, believes in robotics. But he has some reservations. "In a small business like mine," he warns, "you can't just plunk down a couple hundred thousand dollars on a robot that's going to run a job for a week or two. The trick is to be able to develop robotic systems inexpensively that are adaptable across a wide number of operations. Many engineers and technicians can develop a robot easily for a specific application—but to have a multi-use robotic environment is extremely difficult. Factories such as ours need a robot that, after it performs its function, can be moved over to a new spot and begin a second task. When that task is finished, we want to move the robot to a new location to work. We have to have someone who can figure out how to accomplish this with as little downtime as possible."

Another reason small companies have been reluctant to invest is because equipment they have in place may not be used to its fullest potential. In late 1992, *U.S. News & World Report* estimated that U.S. factories were running at less than 80 percent of their capacity. If factories can already produce more than they are selling, they have little incentive to increase capital spending.

These factors—along with tax laws—tend to slow down the rate at which small- to medium-sized companies move toward automation. They affect the numbers of robots purchased and shipped—and, consequently, how many robotics jobs there will be.

TECHNICAL FACTORS

Technical factors, too, play a role in determining whether robots are practical for any given application. One of the virtues of robots—the ability to return to the same spot again and again after a point has already been "taught"—can also be a drawback in their use. For instance, parts that the robot is going to pick up must be oriented and positioned correctly. If the part isn't facing the direction the robot expects and isn't where the robot expects it to be, there may well be problems in the assembly process. Engineers who are considering automation generally try to design (or redesign) parts so they can be fed and inserted correctly. For instance, the Society of Manufacturing Engineers (SME) has reported a redesign of an IBM printer component. The original design had 27 parts; the redesign has just 14.

Details can make or break a project. In the design of automated systems for food processing, a 1989 test gave good results for chickens in one section of the United States. However, when the system was installed in a location in another region, it had to be modified before it would work successfully. The reason? Chickens in the second plant had drier skin and were of a different size from the chickens in the test plant.

ROBOTICS IN THE AUTOMOTIVE INDUSTRY

By late 1992, approximately 44,000 robots were installed in U.S. factories, according to the Robotic Industries Association (RIA), a trade association that tracks robot statistics. "The automotive industry and other heavy manufacturing industries are still the primary customers," says Donald A. Vincent, RIA's executive vice-president. "And the largest application areas for 1992's new robot orders were in spot welding, material handling, arc welding, and assembly."

The University of Michigan Transportation Research Institute has been studying the U.S. automotive industry for a number of years. Its Delphi studies, available for purchase, present the combined opinions of more than 300 authorities who answer questions about technological, marketing, and materials development within the automotive industry. Selected results from Delphi VI, "Forecast and Analyses of the U.S. Automotive Industry through the year 2000," indicate that the overall North American vehicle market will remain stagnant until the turn of the century. In fact, Delphi VI panelists foresee continued Big Three market share erosion in the passenger car and light truck market in the United States and North America. Overall, panelists believe traditional American manufacturers' market share will drop 3 percent between 1990 and 2000 in North America—from 71 percent to 68 percent.

No one is certain just what this will mean for jobs in robotics. However, General Motors spokesman Mark Tanner says that GM plans to reduce the number of employees in North American operations by approximately 74,000 by the mid-nineties. It is possible that automation—and the use of robots—will become a significant way to lower labor costs; it is also possible that current employees will be retrained to troubleshoot and maintain existing robots in automotive plants.

Like other companies in the automotive industry, GM uses robots for arc and resistance welding, for assembly, for dispensing and applying sealants and adhesive, for painting, for inspection of vehicles, for loading and unloading machines, for transferring parts from one conveyor to another, for packaging, and for palletizing.

Robots save money, GM says, because they are so precise and repeatable. The fact that they are consistent makes them attractive. Overall, they can produce parts of consistent quality with higher average quality control than that found on parts manufactured by humans.

Union agreements, common in the automotive industry, generally provide that no worker can be displaced because of automation. And the union isn't necessarily fighting the introduction of robots. "As an organization, the U.A.W. is not opposed in principle to the idea of making jobs easier," says Mickey Long, international representative, U.A.W., Ford National Education Training Center. "Clearly, in certain situations, and I'll use an example, automating some of the foundry jobs, was an improvement, not only for the company, but for the workers themselves. We're not opposed to that."

Often, using a robot instead of a human in hazardous situations or less-than-desirable environments makes sense. Workers who perform arc-welding operations frequently find that wearing the necessary (and required) personal protective equipment can be a nuisance. Substituting arc-welding robots that can average a much higher arc-on-time often results in productivity improvements. Loading punch presses by using robots, rather than humans, helps keep workers safer.

In GM and in many plants of other companies, robots are programmed by engineers. Much of the training is done off-line. An engineer develops a program for the robot while sitting at a computer keyboard, takes the programs to the factory floor, and tries them. The engineer, or perhaps a skilled tradesperson, fine-tunes the program. Sometimes, if a robot breaks down, an alarm sounds, and a skilled tradesperson is dispatched to fix the robot. At other times, a succeeding robot down the line may pick up the tasks of the "dead" robot in addition to its own.

Off-line programming isn't always necessary, though. Pacific Robotics, Inc., has an industrial robot for palletizing and depalletizing applications that can be taught to perform a specific work cycle—by someone who does not even have computer programming experience. "A handheld teach pendant with a few control buttons allows the operator to lead the PR-110 robot through the required moves and actions and to record the data in the computer memory on a point-by-point basis," says Robert L. Prendergast, vice-president, sales. The process can be repeated to store as many patterns in the computer as desired.

The control panel uses an interactive display terminal to communicate with the operator in plain text. To begin automatic operation, the operator simply selects the desired pattern and presses the button marked "start." The computer then automatically coordinates simultaneous movement of all four axes to move the load smoothly and swiftly. Pickup and release of the load is also automatically controlled by the computer. The computer can interface the robot with a broad range of devices: limit switches, optical sensors, solenoid valves, and outside computer bases.

At the Presto Food Products, Inc., food processing plant near Los Angeles, a PR-110 robotic palletizer has been installed on one of the process lines. The robot handles different shipping carton sizes and variable pallet patterns. Plant manager Ray Fryxell, Jr., reports that the robot has increased the capacity of the packaging line by almost half while reducing the labor required to run it by 40 percent. The robot automatically picks up one, two, or three shipping cases containing tubs of whipped topping at a time from the end of a conveyor, swings them 180 degrees to a pallet, and drops them precisely where needed to build the specified stacking pattern.

"The robot can pick up a 110-pound load and place it on a shipping pallet anywhere within a 16-foot diameter semicircle," explains Prendergast. "Having an 8-foot reach on this robot lets us serve several conveyor lines as well as multiple pallet loading stations. Consequently, we can cover a number of different loading requirements with the same machine."

WORKING WITH ROBOTS

What do you need to know if you want to work with robots? And how can a potential employer assess your skills?

One way an employer can determine your qualifications is through using material developed by the National Occupational Competency Testing Institute (NOCTI). NOCTI, established in 1969, is the nation's largest provider of occupational/technical assessments. NOCTI operates data-proven and comprehensive testing services, state-of-the-art computer scoring services, job and task analysis workshops, and national/international consulting services. Recent contracts have included an impressive list of clients: Toyota USA, Nissan USA, Buick-Oldsmobile-Cadillac division of General Motors, Nabisco, Central Oil Coal, the U.S. Department of Labor, the states of Pennsylvania and Connecticut, General Electric/Jamaica, and over 700 schools and local school districts.

In 1988, NOCTI completed a two-year U.S. Department of Labor contract for a project, "Robotic Technician High Technology in-plant Training Model." Today, companies are using the competency tests developed during the project, reports Allan Chrenka, director of the Michigan Occupational Competency Assessment Center. In addition, NOCTI is sharing its expertise with a number of community colleges (primarily in Michigan) as those schools work to develop curricula in automated manufacturing technologies.

At the start of the project, the Advisory Council developed a comprehensive statement about robotic technicians—a statement that included work environment, scope of responsibility, and areas of specialization.

"Within the field of computer integrated manufacturing, companies apply computer science technology to design: planning; production management; financial control; movement of materials, communication and commands; the use of machines and equipment (robot work cells) to fabricate, assemble components, and assemble products; monitoring and controlling quality; warehousing, and the utilization of professional and technical personnel to achieve maximum quality and productivity," the council said.

"The Robotics Technician, in most situations, concentrates on responsibilities involved in robot work cells processes with a CIM system."

As NOCTI sees it, the robotics technician must know and understand concepts of physics and mathematics, electricity, electronics, computer science, and other basic disciplines that underlie advanced technology. He or she will be specialized in one or more areas of electromechanical technology:

- robotics equipment and application
- robotics controls and software
- robotics service and maintenance

A technician who specializes in service and maintenance will need training and on-line experience in programming, installation operation, service, maintenance, and troubleshooting of operational problems.

The Robotics Technician usually works under the supervision of a managerial or manufacturing engineer. However, the level of technician will depend on the size of the manufacturing firm, the nature of the production process, the technical complexity of robot work cells, and the demonstrated competency of the individual technician.

The major goals of the project, which was funded by the U.S. Department of Labor–Employment and Training Administration, were to:

- Identify criteria which will be useful for selecting employees who will be successful in training in the field of robotic technician.
- Establish a course curriculum and training format which can be used for the in-plant industrial setting. The training format should be useful in objectively evaluating the various skill levels of trainees, both before and after training.
- Develop the measures to objectively test the competency of instructors in an industrial training setting.
- Identify procedures for the award of academic credit or advanced standing for those who have successfully completed high technology in-plant training programs.

As a result of the project, NOCTI developed an Automated Manufacturing Systems Technician Test Item Bank and Specifications, available on computer disk in ASCII format. The item bank has two assessment categories: an Industrial Automation Assessment, which evaluates the level of knowledge and competency needed for a trainee or student to advance in the field, and an Automation/Robotics Item Bank, a "job-ready" item bank and the journeyman equivalent level, based on industry specifications.

The industrial automation assessment evaluates such items as engineering graphics, statics and dynamics, design of machine elements, industrial manufacturing, and safety in the modern facility. The automation/robotics item bank evaluates integration and teaching of assemblies of integrated system units (ISU). Additionally, for ISUs, it evaluates operation and preproduction implementation, preventive maintenance, scheduled maintenance, immediate/emergency service, modification of service applications, maintenance and storage of documentation, and continuing education.

To purchase a copy of the robotic technician project final report or the item bank package and to learn more about NOCTI, write National Occupational Competency Testing Institute, Ferris State University, 409 Bishop Hall, 1346 Cramer Circle, Big Rapids, Michigan 49307.

ROBOT-RELATED JOBS

There is more to a robotics-related job than just programming or servicing a robot. Nick Ihnat works for Jergens, Inc., a manufacturer of tooling components. "The products we import are pick-and-place robots," he says. "They have a dedicated pickup point and a drop-off point. They load and unload material into machines.

"My company offers accessories: grippers, pneumatic slides, and conveyors." The industries Jergens Inc. Supply typically deals with include the automotive and electronic industries. Customers include GM, Chrysler, Ford, Sony Digital Audio, and Capital Records. "Anything," Ihnat says, "where they have mass production and can set up dedicated equipment to turn out products."

At his Cleveland-based company, he says jokingly, management considers him "the automation specialist." As an outside sales rep, he has the whole continental United States as his territory for the robotics line of end-effectors. He oversees sales representatives in the field, giving them aid and technical support.

Though Ihnat is just 28, he has been with the company 13 years—the first eight spent in shipping and receiving. After he graduated from high school, he put himself through Lakeland Community College. He is a firm believer in co-op programs and says persons who get experience will go a lot further than those who just have "book-learning."

Ihnat understands the technology his company sells, but he goes a step further. He gets his company's competitors' catalogs and reads them—"front to back." He compares their products to those he sells. He does his homework. He matches what his company offers to his customers' needs. Starting salaries for a job like his, he says, are around $20,000; today, as a veteran, he receives salary plus commission on sales.

Automation and its components will grow, he believes. "Whether you're assembling electronics in a clean room environment or casting steels in a foundry to make engine parts, labor costs are going up, up, and up. Companies today are demanding quality products. Machines (including robots) make fewer errors than humans."

CHAPTER 6

MACHINE VISION, SIMULATION, AND ARTIFICIAL INTELLIGENCE

Three important technologies are related to robotics: machine vision, simulation, and artificial intelligence. Scientists and researchers are studying each of them. The technologies are important in helping people to develop robots with more capabilities or to help those who want to use robots decide what will work best.

MACHINE VISION

When you add sensors to a robot, you give it the ability to tell the computer that controls it what the robot "sees" or "touches." Of course the robot does not really see. It measures changes in conditions and responds to those changes, much as a human worker would.

"Robotics addresses a remarkable spectrum of systems concepts, domains, and applications," says Paul Schenker, senior scientist at NASA's Jet Propulsion Laboratory and editor of *Robotics,* a newsletter published by SPIE—The International Society for Optical Engineering. "Sensors are ultimately rooted in decision and control systems and their implication, biological and artificial. Robotics makes this sensor/motor relationship very explicit. Robotics provides challenges in sensor-based planning, active sensor control, visually guided navigation, multisensor fusion and allocation, and human-sensor interface in man-machine systems."

While machine vision (one of the common sensing technologies) is a technology used with robots, only about 20 percent of the machine vision systems in use in the United States represent a tie-in with the robot industry, according to Donald A. Vincent, executive vice-president of the Robotic Industries Association. In fact, the growth of machine vision systems applications has apparently surpassed that of robotic manipulators.

Dr. Kevin G. Harding, principal scientist at the Ann Arbor, Michigan-based Industrial Technology Institute, says that today almost every major manufacturer has an in-house group with some expertise in machine vision application. And Market Intelligence Research Corporation, a company that tracks the machine vision world market, predicts that world machine vision use will double by 1998. The corporation believes that sales will increase from $478 million in 1992 to over $1 billion by 1998—a projected 12 percent compound annual growth rate.

What Machine Vision Systems Can Do

In general terms, machine vision can be described as the capability to sense, store, and reconstruct a graphic image. But in practical terms, machine vision systems usually have a specific task. They can check to see that parts are oriented properly. They can identify parts. They can search for specific defects or can check that parts are aligned for assembly. And they can do much more.

For instance, a manufacturer of pregnancy testing kits that are used at home had to automate the assembly of the kit's components while still meeting standards. Each kit must be inspected to be sure all parts are present and are placed properly. In addition, the manufacturer has to verify a minimum value for one of the kit's components by checking color. A three-camera vision system solved the problem by inspecting four of the kit's components as it was being assembled. The system can tell when assemblies are bad and will remove them from the production line.

Another manufacturer—this one, in the automotive industry— needed an inspection system that could sort styles of die-cast aluminum wheels. At the plant, up to twenty casting machines produce different wheel styles. All of them feed onto a single conveyor. But some of the wheel styles must be taken off the conveyor for heat treatment. Purchase specifications for the vision system included the system's ability to learn up to 40 wheel styles and to make virtually no mistakes in classifying them correctly.

The turnkey machine vision system that the manufacturer purchased uses a high resolution camera. When a wheel passes under an inspection chamber, the vision system "sees" and "remembers" its image. Then the vision system inspects the image it has captured in its memory. Once the wheel is identified, the vision system searches its internal database to find where the particular wheel style should be sent. The system signals the routing direction, and the wheel is diverted. Each part processed is logged electronically into a report that contains the number of each of the wheel styles processed.

Other machine vision systems inspect crowded printed circuit boards, making sure they are soldered uniformly and properly. Some vision systems check pharmaceutical products, making sure they are labeled correctly. Still others are used to search carry-on luggage at airport security checkpoints. They look for plastic explosives and suspicious organic substances—drugs, cash, gems, and food. Similar systems are used to analyze furniture as it is moved in and out of high-security buildings. In fact, it is possible to search entire trucks, trailers, train cars, and sea cargo containers automatically for unauthorized goods.

How Machine Vision Works

The first step in a machine vision system is to gather data from the original image. Generally, a video camera takes a picture of what it sees. Then the information is fed electronically into the computer that is part of the system. The computer processes the information into a matrix of digitized dots (called pixels). Next, the computer analyzes the pixels for patterns. It quickly searches the huge dictionary of possible objects that has previously been programmed into its memory.

Since the computer can process billions of bits of information per second, the vision system can decide whether what it is seeing is acceptable or nonacceptable" according to the standards that have been previously set. The system—or the robot that may be using the system—can accept or reject a part or component.

Some vision applications require higher quality than others. *Resolution*—the ability of the scanning system to distinguish between two closely separated points—affects the quality of the image the system produces. So does *contrast*—the ability of the scanning system to detect shades of difference from pixel to pixel.

Simple vision systems register each pixel as either black or white. More sophisticated vision systems can detect and reproduce up to eight shades of gray (from black to white) for each pixel.

Sometimes, though, simpler systems are better. If the vision system is looking for the edge of a part, it doesn't care whether that edge is gray, black, or white. The system just needs to be able to find the edge—time after time, within a specified range of accuracy.

Lighting, too, is an important component of successful machine vision applications. Because humans and computers don't "see" things in the same way, machine vision systems may perform better with different lighting systems from those human inspectors would prefer. All lighting systems do not produce the same images. Back lighting, for example, can't easily light parts that are moving on a conveyor belt unless the belt is made of a clear or translucent material. Dark-field illumination (a form of back lighting) shows imperfections as bright spots in an otherwise dark field. Side lighting can create shadows that will highlight electronic chips on a board. Diffuse lighting that comes from a distance at multiple angles makes all shadows disappear.

Although the sensor that gathers the image for a machine vision typically is an *analog* device, the signals must be *digitized* in order for the computer to store and analyze them. You can understand these terms more easily if you think about how objects are measured.

For instance, if you consider how a young child grows, you know that the child becomes taller, imperceptibly, on a day-by-day basis. Even if you use a yardstick and try your best to measure accurately, you can't say a toddler is exactly $36^{1}/_{2}$ inches tall on Monday morning and $36^{52}/_{100}$ inches tall by Monday afternoon. Analog signals (like those used in the original telephone system) are continuous variables. Digital signals, on the other hand, are signals encoded as a series of discrete numbers. If you think about a partly filled carton of eggs, you have four eggs or five eggs or seven eggs. You don't have four-and-a-half eggs. Computers "understand" signals in the same way you understand the number of eggs in the carton.

After the image has been scanned and stored in the computer of the machine vision, the system analyzes the image. It uses a number of methods to do so, and—if desired—it can identify the image by its shape.

Some vision systems make sure a package is lined up correctly to be labeled or that bottles are filled to the right level. Others are used to measure car bodies. General Motors has a gauging station at one of its plants that uses more than a hundred cameras to record measurements of gaps and contours in metal parts in just 20 seconds. At Ford Motor Company, one system identifies types of car bodies, and another one inspects bolts used for assembling cylinder blocks. Austin Rover uses machine vision to align and place windscreens on cars, while another car manufacturer inspects brake assemblies and painted surfaces. Machine vision performs precise measurement for part remanufacture. It inspects the mastic on automobile doors and checks pistons and tests speedometers to guarantee quality and accuracy.

Green Giant has used machine vision to inspect and measure corn cobs, and Kodak inspects photographic film. Other companies inspect glass containers and woven belts.

SOURCES OF INFORMATION

Society of Manufacturing Engineers

There are several places you can write if you want to learn more about machine vision and its applications. You may want to order a copy of the latest issue of *Directory of Manufacturing Education,* available from the Society of Manufacturing Engineers, One SME Drive, P.O. Box 930, Dearborn, Michigan 48121-0930. The book includes colleges and universities with both machine courses and equipment at the associate and bachelor-degree level and at the master's and Ph.D. level.

Automated Imaging Association

Another good source is the Automated Imaging Association (AIA). This association promotes the acceptance and productive use of image processing, image analysis, and machine vision technology. Members are manufacturers of related and peripheral products, integrators, end uses, consultants, and research groups.

AIA has several activities primarily for members, but in general, students or interested persons can attend some of the sessions and view exhibits if they pay appropriate fees. Among them: the International Robots and Vision Automation Show and Conference and the Machine Vision and Imaging Financial Forum. At the former, more than 200 companies exhibit automation solutions for manufacturing while more than 100 inter national experts give conference presentations. At the Financial Forum, industry executives join leading financial analysts and venture capitalists to talk about the outlook for the industry.

For information on the AIA, write the Automated Imaging Association, 900 Victors Way, P.O. Box 3724, Ann Arbor, Michigan 48106.

Machine Vision Association

Another source of information on this technology is the Machine Vision Association of the Society of Manufacturing Engineers (MVA/SME). Founded in 1984, MVA/SME has over 3,300 members in 35 countries. It publishes books and conference proceedings and has produced several videotapes. In addition, it offers a number of reference materials for sale, including the *Machine Vision Industrial Directory,* a listing of principal suppliers of vision systems, components, and services; a *Bibliography of Machine Vision Technical Resources,* a guide to every technical paper and article published by SME on machine vision; and *Roundtable Transcripts,* which features industry experts discussing timely issues of concern to manufacturers and users of machine vision systems.

For information on MVA/SME, write to the Society of Manufacturing Engineers, One SME Drive, P.O. Box 930, Dearborn, Michigan 48121-9939.

International Society for Optical Engineering

Another association to contact is SPIE—the International Society for Optical Engineering, which, in 1992, established an International Technical Working Group on Robotics. "Sensors and signal processing play many roles in robotics design and applications," says Paul Schenker, chairman of SPIE's Robotics Working Group, senior scientist at NASA's Jet Propulsion Laboratory, and editor of *Robotics,* the technical working group's newsletter.

For information, write to SPIE—the International Society for Optical Engineering, P.O. Box 10, Bellingham, Washington 98227-0010.

ROBOTIC SIMULATION

Simulation—trying out a process before actually installing it—can be a money-saver in robotics. Factories considering installing a robot have many questions to consider: how many robots are needed, what kinds of robots are best, and how the robotic system can best be placed on the plant floor for an efficient layout.

One way to design and analyze a robotic work cell is to simulate the motion of its components—to rotate and visualize a three-dimensional model. Such a model actually mathematically represents the robot and its components. Using simulation, the robot's operations can be programmed so the factory engineers can see what works and what doesn't.

C. Ray Asfahl, author of *Robots and Manufacturing Automation,* tells the story of the testing of a robotics application. As he remembers, the Singer Company's Motor Products Division was installing a new robot—the first in the plant. "A few seconds into the test," Asfahl says, "and crash: a sequencing error caused a collision between the robot and its simulated machine test stand, which literally fell over." Fortunately, the test was a simulation, and simulated machines had been substituted for real equipment.

Currently, robotic graphics and simulation are being used in a variety of manufacturing processes, including welding, assembly, press operations, palletizing and packaging, machine tending, moving assembly lines, and tool handling.

Simulation technology, of course, has a number of other applications besides robotics. As defined by John McLeod, founder of the Society for Computer Simulation International (SCS), simulation is the development and use of models that help evaluate ideas and study dynamic systems or situations. "Computer models mimic the operation of the characteristics of a system," he says. "And computer simulation is the use of a computer model to study 'what if?' situations."

A visual simulator—a device that uses computer-generated imagery to create highly realistic, completely unrestricted environments—has a number of key elements. Among them:

- A three-dimensional computer model (called a visual data base). Just like a list that has been entered in a personal computer's database program, the visual data base is made up of elements that can be manipulated and rearranged.
- A computer-image generator that creates and renders the simulated scene in real time—right then and there.
- A host computer that supervises overall system operation and manages the dynamics of the simulated operation.
- A monitor or display system that shows the images that the computer-image generator has created.

Visual simulators are often used in industry to train operators. For instance, at Florida Power Corporation, Merrill Quintrell, senior fossil operations training specialist for the coal-, oil-, and gas-burning utility, uses them for plant-specific training on emergency management. "Our task force found that operator error was part of virtually all our damaging boiler incidents and near misses," he recalls. "We needed to train and retrain."

Quintrell, who has a background in computers, plant operations, and training, wrote specs, oversaw construction of a specially built simulator, took delivery, made sure everything worked, and customized the simulation software. But he doesn't rely on simulation technology alone to get the job done.

"After my students have reviewed their plant emergency procedures, I cover all the possible malfunctions," he explains. "They never know in advance which malfunction they'll get when they're in the simulator phase of training." Then he uses a chalkboard in the simulator room to discuss what they have done right—or wrong—and talks about how the laws of physics apply to the particular emergency they have handled.

Top-of-the-line simulators aren't cheap. Gene Yokomizo, United Airlines manager of flight simulator services, estimates that the company's 26 fleet-specific simulators cost between $11 and $20 million—each. However, all of United's 8,000 pilots spend three days in simulator training as part of an annual proficiency check at the airline's flight center at Denver's Stapleton International Airport. In addition, pilots who change from one type of aircraft to another are retrained on the simulator for two weeks of a mandatory 30-day period. Pilots who upgrade their status—for example, a promotion from first officer to captain— also get two weeks of simulator time as part of their required 30 calendar days of training at the flight center.

ARTIFICIAL INTELLIGENCE

Artificial intelligence is a technology that applies to computer systems and, by extension, to robots. An "intelligent" computer system shows the characteristics we generally associate with intelligence in human behavior, such as understanding language, learning, reasoning, and solving problems.

In short, artificial intelligence (often called AI) is the way in which a machine is manipulated by a human to do a human function. Or, as Harvey Newquist defines it for *ComputerWorld,* "AI is a group of technologies that attempt to emulate certain aspects of human behavior, such as reasoning and communicating, as well as to mimic biological senses, including seeing and hearing.

"Specific technologies include expert systems (also called knowledge-based systems), natural language, neural networks, machine translation, and speech recognition."

It may seem difficult to make a computer or a robot understand language. Yet Lockheed has been using voice-data entry (speaking to a machine and telling it what to remember) for a number of years. Using artificial intelligence, computers have successfully played difficult games, understood simple sentences, performed useful industrial work, and even exhibited learning behavior.

Artificial intelligence does not mean that a computer (or robot) "thinks" by itself. Instead, it has been programmed to process information in certain ways.

Sometimes this information can be acquired by experience the machine gains. For instance, a computer can be told, "If the power supply to the radar on board fails and a backup power supply is available and the reason for the first failure no longer exists, then switch to the backup supply." If the computer has been given those rules beforehand and the capability to sense the conditions under which those rules should be applied, it can follow those instructions by itself.

A robot that can be programmed to choose how it will perform or what tasks it will do, depending upon the input it receives from its sensors, is called an intelligent robot. But by themselves, robots have limitations. Unless they are equipped with sensors, they generally cannot make decisions based on their environments. They have a limited capability.

Some applications require an intelligent robot. Others don't. As C. Ray Asfahl, author of *Robots and Manufacturing Automation*, explains, users must weigh trade-offs. Often, they must decide what is more important in a particular application. A common example may be speed versus sophisticated analysis in a machine vision system. If the system needs to accomplish several objectives and if artificial intelligence is required, the system generally will slow down. Perhaps that is acceptable; perhaps it is not. Or, if a manufacturer is bringing the first robot ever into a plant, the company may not want to start with a sophisticated sensing system and artificial intelligence. Instead, Asfahl recommends using the simplest robot that will accomplish the desired objective.

Like simulation, artificial intelligence technologies are used in fields other than robotics. In fact, more than 13,000 members from various disciplines belong to the American Association for Artificial Intelligence (AAAI), the principal technical society serving the Artificial Intelligence community in the United States.

Founded in 1979, the association has members who are business planners, corporate managers, industrial research scientists, knowledge engineers, computer science professionals, psychologists, linguists, cognitive scientists, academicians in related fields, and engineers. Subgroups have been established in medicine, manufacturing, law, and business.

The association sponsors a major annual conference: the National Conference on Artificial Intelligence. Proceedings, technical programs, and exhibit programs are published. Case histories of working applications are featured in the AI On-Line Series.

Issues like management involvement, maintenance and service-ability, system integration, and design are discussed.

In addition, AAAI sponsors spring and fall symposium series—smaller forums for discussions on advanced subfields within AI. The fall 1991 series included symposia on Hybrid Reasoning and Sensory Aspects of Robotic Intelligence.

Tutorials, which focus on technology transfer, are attended by representatives of business, research laboratories, and academic institutions. They are divided into three groups: Emerging Technologies, Industrial Applications (which focus on a specific application area), and Practical Issues (which focus on systems building and management issues).

Currently, artificial intelligence systems are being offered for applications in manufacturing. They are controlling and monitoring automated material handling systems. They are helping factory managers design, plan, and control flexible manufacturing cells.

As research in artificial intelligence continues, robots that can use such knowledge and processing techniques can become even more valuable. You can learn more about artificial intelligence and the exciting challenges the technology offers by reading back issues of *AI Magazine*. They can be purchased from the Publications Department, American Association for Artificial Intelligence, 445 Burgess Drive, Menlo Park, California 94025-3496.

CHAPTER 7

NONINDUSTRIAL (SERVICE) ROBOTS

Not all robots are industrial robots. Growing numbers of robots are used in education, health care, security, training, space, and military operations. Within Robotic Industries Association, a specialty association called the National Service Robot Association (NSRA) represents builders, developers, and users concerned with this application of technology.

"The service robotics industry has the potential to be a very large and important industry in the days ahead," says Jack O'Brien, a veteran of 25 years at Polaroid, who has been responsible there for developing and implementing the company's strategies for sensor development. Polaroid's ultrasonic sensors are often used in mobile robots.

O'Brien believes that key applications for service robots in the coming years are in health care, security, space and underseas exploration, construction, food service, nuclear maintenance and clean-up, and education. Researchers are looking at additional applications. For instance, in the United Kingdom, the Department of Trade and Industry's robotics program targets six priority service robot areas: tunneling, construction, underwater, firefighting and security, medical, and domestic.

Service robots are being developed for applications you might never have imagined. Scientists at the University of Washington are working on a mobile robot to handle luggage at airports. Their informal name for the project: Roboschlepper. The National Institute for Standards and Technology (a U.S. government-funded agency) has developed a robotic crane they call the NIST Spider. Potential applications in the service industry include excavating and grading. And Tennessee Tech's Center for Manufacturing Research and Technology Utilization is working on a tower-climbing robot to paint rusting utility towers.

On the medical side, the Applied Science and Engineering Laboratories of the Alfred I. duPont Institute are studying ways to bring robotic and advanced technology into applications for persons with disabilities. They are concentrating their research on the interface between the human and the machine. A California-based company has petitioned the U.S. Food and Drug Administration, asking it to approve human hip-replacement surgery using the "Robodoc." The robot has already helped perform 25 hip replacements in dogs.

WHY USE SERVICE ROBOTS?

Companies that use service robots see them as a cost-effective way of accomplishing tasks. For instance, the health care industry is looking for ways to decrease costs while still maintaining—or even improving—the quality of patient care. Labor costs, especially when benefit costs are factored in, are high. Consequently, hospitals, clinics, and nursing homes are looking at all internal operations, seeking areas where productivity can be increased.

Among other issues, they are studying ways to improve "fetch-and-carry" tasks. It is inefficient, health care administrators feel, to use highly skilled (and highly paid) hospital personnel for these tasks. As a solution, Transitions Research Corporation, a Connecticut-based supplier, developed HelpMate, a highly sophisticated mobile robot. HelpMate, used in hospitals in Danbury, Connecticut, and Downey, California, is saving money while increasing productivity. "We've been having to pay someone a minimum of $6.65 an hour to go from point A to point B," says a Downey Hospital administrator. "Add to that all of the other normal payroll costs, and it makes sense to invest in a piece of equipment like the robot."

A reduction in workers' hours—with a corresponding saving in costs—is what Danbury Hospital sees as the prime benefit for robot use. "The kitchen has reduced the number of hours we schedule per week from 1,015 to 975," a Danbury administrator says. "The 40-hour savings is a direct result of using the robot."

One of the potential areas for growth in the field of service robots is their use in the cleanup of hazardous waste. Military installations and defense contractors dumped waste byproducts at a number of sites during the defense buildup in the Cold War period. "At some of those sites, no one is quite sure just what is buried," says Don Vincent, executive vice-president of the Robotics Industries Association. "OSHA and EPA regulations—and common sense—mean you can't just send humans in there to clean up, since you don't know the exact hazards involved. Service robots can help." Westinghouse already uses service robots inside nuclear power plants to clean and repair pipes. The robots can do the job more safely (and faster) than people, saving plant downtime.

ROBOTS IN SPACE

Business Week has reported that Martin Marietta Corporation is developing a robot that would put together and maintain a space station. But such research is not new. NASA has been studying robot research and development, primarily at its Pasadena, California, Jet Propulsion Laboratory (JPL), for over 10 years. NASA is looking at two areas:

- Systems autonomy research, which concentrates on applying tools of artificial intelligence to mission planning—tools such as expert systems, planning aids, full diagnosis capabilities, and relational data bases.
- *Telerobotics* research which looks at servicing, assembling, and manipulating the capabilities of humans in a space environment through the use of robotics.

Systems autonomy is necessary in space exploration because the distances between the planets are so great that it takes many minutes for humans on Earth to communicate with spacecraft and planetary landers. A remote machine—a robot—that is capable of making independent decisions, however, can accomplish more on its own without needing step-by-step instructions from its human supervisors.

Telerobotics, however, means teleoperator plus robot, according to Dr. Paul Schenker, robot expert and editor of the SPIE *Robotics* newsletter. "A teleoperator is a human being who controls a humanlike set of manipulators and effectors to observe conditions and carry out a manipulation task," he explains. "Usually the human is in one location, and the manipulators and effectors are at another site—often, a remote one."

Schenker's description brings up an important point you should consider as you explore opportunities in robotics careers. Virtually all service robots still must be controlled by a human operator. Unlike robots on a factory floor, service robots must work in variable environments. Because the conditions they operate in can change quickly, because the tasks they are called upon to do vary, service robots cannot yet be programmed to handle the necessary complexity. In short, your chances of working in robotics may be greater in this area than in industrial applications if the service robot field lives up to its promise. Many of the service robot applications and technology are being developed through laboratory research. Write to the Society of Manufacturing Engineers, One SME Drive, P.O. Box 930, Dearborn, Michigan 48121 to purchase their Robotics Research Directory. The book describes the various university-connected laboratories where research on robotics is being conducted; write to the addresses listed for details on projects.

CHAPTER 8

EDUCATION AND TRAINING

What do you need to get a job in robotics? Where should you go to school? What should you study?

There are no simple answers to these questions and no one formula for success. But increasingly, experts are recommending a series of core competencies for those who want to succeed in tomorrow's workplace.

SCANS

In 1991, the Secretary's Commission on Achieving Necessary Skills (SCANS) was appointed by the U.S. Secretary of Labor to determine the skills people need to succeed in the world of work. The commission's purpose: to encourage a high-performance economy, characterized by high-skill, high-wage employment. Its members, distinguished leaders from education, business, labor, and government, were charged with defining a common core of skills that constitute work readiness for the jobs of today and tomorrow.

SCANS reports have identified "workplace know-how"— made up of five workplace competencies and a three-part foundation of skills and personal qualities that are needed for solid job performance.

Workplace Competencies

Here are the workplace competencies SCANS says effective workers can productively use:

- Resources—Workers know how to allocate time, money, materials, space, and staff.
- Interpersonal skills—Workers can work on teams, teach others, serve customers, lead, negotiate, and work well with people from culturally diverse back grounds.
- Information—Workers can acquire and evaluate data, organize and maintain files, interpret and communicate, and use computers to process information.
- Systems—Workers understand social, organizational, and technological systems. They can monitor and correct performance, and they can design or improve systems.
- Technology—Workers can select equipment and tools, apply technology to specific tasks, and maintain and troubleshoot equipment.

Foundation Skills

Here are the foundation skills SCANS says competent workers in the high-performance workplace need:

- Basic skills—reading, writing, arithmetic and mathematics, speaking, and listening.
- Thinking skills—the ability to learn, to reason, to think creatively, to make decisions, and to solve problems.
- Personal qualities—individual responsibility, self-esteem and self-management, sociability, and integrity.

Success with Workplace Know-How

This combination of workplace competencies and foundation skills—SCANS calls it "workplace know-how"—is not taught in many schools or required for most diplomas. Nevertheless, your chances of succeeding in robotics or a related career are better if you can perform the tasks identified in the SCANS reports.

A high school diploma was once a sure ticket to a job; today, however, the market value of a high school diploma is falling. The SCANS report *Learning a Living: A Blueprint for High Performance* says the proportion of men between the ages of 25 and 54 with high school diplomas who earn less than enough to support a family of four above the poverty line is growing alarmingly. In 1989, more than two in five African-American men, one in three Hispanic men, and one in five white men—all with high school diplomas—did not earn enough to lift a family of four above poverty. Unless there was a second wage earner in these families, they did not have what most would call a decent living.

The workplace know-how SCANS has defined is related both to competent performance and to higher earnings for the people who possess it. When the know-how required in 23 high-wage jobs is compared with the requirements of 23 low-wage jobs, SCANS says, workers with more know-how command a higher wage—on average, 58 percent, or $11,200 a year, higher.

JOB AVAILABILITY IN ROBOTICS

If you do have the SCANS competencies, if you do have a degree or advanced course work, if you do have factory experience, can you get a job in robotics? Maybe. Maybe not. It is extremely possible that there may currently be more people trained in robotics than there are jobs to fill.

In 1980, there were about 20 schools and colleges offering courses in robotics; by 1987, there were over 400; by 1992, there were even more, although many of the courses offered fell within broader, manufacturing- or engineering-based curricula. Yet figures from Robotic Industries Association, the national grade group for vendors and suppliers of components, show that only 2,600 robots were ordered from U.S.-based robotics companies during the first half of 1992—the highest first-half total since 1986. RIA estimates that some 44,000 robots are installed in U.S. factories—total—as contrasted with Japan, where some 50,000 robots are installed *each year.*

Does this mean your chances for a job in robotics or related technology are hopeless? Not at all. It does mean that you may be disappointed if you go to school specifically to study robotics, take all the robotics-related classes offered in the curriculum of the school you are attending, and expect to have a for-sure job in robotics waiting when you graduate.

ACTIVELY PLANNING YOUR ROBOTICS CAREER

If you want to enter robotics, you are going to have to be extremely active and alert to recognize industry trends and to plan your career accordingly. There are several reasons this is so. One significant problem is that many young people and many high school guidance counselors have little or no firsthand experience with modern manufacturing technology. Unless you live in a community with factories that use robots, you may never have seen an automated production line. Unless your school counselor has toured such factories, he or she may not know the opportunities robotics as a career has to offer—or its limitations.

Another reason you must take an active role in planning your training is that the robotics market is changing rapidly and is influenced by many factors unrelated to courses in school. Major companies like General Motors are downsizing in order to remain competitive and viable in a rapidly changing global marketplace. GM spokesman Mark Tanner says the corporation is reducing the number of employees in its North American operations by the mid-nineties—down 74,000 workers from employment levels in December 1991.

Downsizing this severe—and GM is not the only manufacturer making cuts—has a ripple effect that changes the employment picture for robot manufacturers and suppliers of components. Offshore sourcing—in which companies put together parts and components overseas, where production costs are lower—has also become a significant factor in manufacturing.

Still another factor affecting your career in robotics is an agreement many companies may have with unionized workers, promising that jobs will not be lost because of automation. Those workers already employed may need retraining, but they often get first chance at jobs. As an inexperienced worker, you are not as valuable as a veteran employee who already knows the company and the production process.

Don't be discouraged, however. There are a number of steps you can take to improve your chances of finding work in robotics.

Skills You Need

If you are still in junior high or high school, take all the math, science, and computer courses you possibly can. Today, being in robotics means having a technical background. "People nowadays who think they can get anywhere without a strong math and science background are fooling themselves," says one expert at the Jet Propulsion Laboratory's Technology and Space Program Development Office. Because robots are controlled by computers, an awareness of programming and familiarity with computers are essentials, JPL experts say.

Communications skills, also, are seen as a must by many experts. You are almost certainly going to have to make reports on what you are doing; consequently, you must be able to speak and write well.

Career Days

If you are fortunate enough to live in or near an industrialized area that has automated factories, you can suggest that a person who works with robots be invited as a career day speaker. That way, you will be able to ask questions about job opportunities and limitations. You can find out whether factory tours are available and whether the factory hires students for summer or part-time jobs. At least you can establish your interest in robots early.

Video Programs

Several award-winning video programs are available free of charge to schools, libraries, and career counselors. *Challenge of Manufacturing*, intended for junior or senior high school students ages 12-19, features people working as manufacturing engineers producing bicycles, blue jeans, compact discs, and cosmetics. *Race Against Time*, designed for persons ages 15 and up, provides information for people evaluating manufacturing as a career path. The program explores the status of manufacturing, new strategies in use, competition, teamwork concept, and the rewards of manufacturing. *Engineering: Making It Work* emphasizes how manufacturing touches our lives in everything we do, from music to make-up, cars and clothing, jet fighters, and rock and roll. It's suitable for all ages and grade levels.

While none of these videos specifically targets robotics as a career, nevertheless, each shows the factory environment and discusses activities young people can do while they are still in high school to help them prepare for manufacturing engineering. Career brochures and posters can also be purchased for use with the videos. They promote manufacturing education to teachers, students, and professionals.

Contact the SME Education Department, One SME Drive, Dearborn, Michigan 48121 for free-loan or purchase information.

CHOOSING A SCHOOL

A *Directory of Manufacturing Education* can be purchased from the Education Department, Society of Manufacturing Engineers, One SME Drive, P.O. Box 930, Dearborn, Michigan 48121. Included in the comprehensive directory is information on over 550 colleges, universities, and technical institutes that have degree programs in manufacturing and related areas—including robotics. Listings include degrees and course offerings; cooperative education and evening programs; and complete names, addresses, and phone numbers of persons to contact.

ABET

If you are planning to take engineering and engineering-related courses as a foundation for robotics studies, you will want to know about the Accreditation Board for Engineering and Technology, Inc. (ABET). For over 60 years, ABET has been monitoring, evaluating, and certifying the quality of engineering and engineering-related education in colleges and universities in the United States. ABET develops accreditation policies and criteria. The board runs a comprehensive program that evaluates engineering and engineering technology degree programs. Programs that meet ABET's criteria are granted accredited status.

The U.S. Department of Education formally recognizes ABET's exclusive jurisdiction for accrediting engineering, engineering technology, and engineering-related education. In addition, state licensing authorities—either by specific statute or by long-standing practice—generally recognize ABET-accredited engineering programs for full educational credit toward satisfaction of state professional engineer licensing requirements. Graduates of ABET-accredited programs have a high degree of job mobility because of the wide recognition of the accreditation system in the world engineering community.

ABET defines *engineering* as the profession in which a knowledge of the mathematical and natural sciences gained by study, experience, and practice is applied with judgment to develop ways to utilize, economically, the materials and forces of nature for the benefit of humankind.

ABET defines *engineering technology* as that part of the technological field which requires the application of scientific and engineering knowledge and methods combined with technical skills in support of engineering activities; it lies in the occupational spectrum between the craftsman and the engineer at the end of the spectrum closest to the engineer.

ABET defines *engineering-related programs* in higher technical education as mathematics- and science-based programs that do not fit the strict definitions of either engineering or engineering technology but have close practical and academic ties with engineering. With appropriate participation from societies representing specific engineering-related professional disciplines, engineering-related programs may be structured to prepare graduates for professional practice in a discipline that is neither engineering nor engineering technology (for example, surveying and mapping or industrial hygiene).

Each year, ABET issues its annual list of accredited programs. In addition, ABET publications, which can be ordered for a fee, include an annual report and publications on engineering accreditation, engineering technology accreditation, engineering ethics, professional development, continuing technological education, and accreditation of engineering-related programs.

A 10-minute cassette videotape is available from ABET. "Engineering and Engineering Technology—Accreditation for Quality Education" provides a description of ABET's accreditation process and the benefits of accreditation to institutions, students, employers, and the engineering profession.

For further information, write to ABET, Publications Office, 345 East 47th Street, New York, New York 10017-2397.

Questions to Ask

Robotics experts suggest you ask several key questions as you visit campuses and select a school.

- Do you have a robotics laboratory? Can I get hands-on experience with robots?
- Do you have an advisory committee for the robotics program with members who are working in robotics? Programs that do are aware of industry trends and of what courses may be needed to train students as employment opportunities change. Some colleges and universities may use part-time faculty who not only teach robotics courses but who work in industry.
- Do you have an engineering co-op program? Schools and universities that let students alternate classes with on-the-job experience may offer more opportunities for students to get ahead. Although it may take you longer to finish your training, you will have practical experience that potential employers value highly. You have proved that you can do a job well.

- Are students prepared for the Society of Manufacturing Engineers (SME) certification exam? Do students actually take the exam, either before graduation or shortly thereafter? What percentage of them earn certification?
- How many students who have recently graduated have gotten jobs in robotics? How many alumni are working in robotics or related technologies? The school should be able to put you in touch with one or two alumni in the field. Or you may want to contact several graduates of the robotics program to discuss their job-hunting and hiring experiences.

Specialized versus Broader Courses

To what extent should you specialize in robotics? Or, instead, should you be studying the broader field of manufacturing technology? Should you be looking at robot repair and maintenance as a viable choice? Or do you need the skills that will help you get a job in factory automation?

Increasingly, those knowledgeable in robotics and manufacturing technology are suggesting your preparation be as broadly based as possible. If employers have workers who could be retrained to service robots, those employees may well be given preference over new hires. Of course, existing employees will need reading levels and math skills sufficient to be retrained in hydraulics and pneumatics.

Companies wanting to automate are not going to turn you loose on an expensive robot if you are fresh out of college. But if you can tell them, "I have hands-on experience with automated manufacturing, including robotics, electronics, computer science, and good math and communication skills," you may be hired to help implement automation—especially in a smaller plant.

WHERE TO FIND COURSES

Schools and colleges offering course work in robotics—usually, as part of a larger offering in manufacturing technology—aren't hard to locate. Here is a sampling.

College of DuPage. This community college provides education in robotics and automated manufacturing. Within its occupational and vocation division, the college offers two programs: one in electro-mechanical and the other in manufacturing technology, which lead to the associate of applied science degree or to one of the numerous certificates.

Course work stresses practical applications in robotics and other technology areas, with extensive hands-on laboratory experience. The lab provides experience in robotics applications, software, hardware, vision systems, and part inspection as well as artificial intelligence. Programs also contain courses in process control, programmable controllers, drafting, numerical control, and CAD/CAM technology. Co-op and job training programs are also available. For more information, contact College of DuPage, 22d Street and Lambert Road, Glen Ellyn, Illinois 60137-6599.

Eastern Illinois University. Robotics is included in the manufacturing technology option of the industrial technology degree program in the school of technology. This option is designed to prepare individuals to meet the challenge of the modern domestic manufacturing industry. The school's approach is to include robotics as part of the study of a larger system of manufacturing.

The focus of the manufacturing technology option is on state-of-the art manufacturing systems that include computer-aided design (CAD), computer-aided manufacturing (CAM), computer integrated manufacturing (CIM), and robotics. Courses in these areas in combination with other management and engineering-related courses, such as manufacturing management, plant layout and material handling, statistical quality control, machine design, work measurement and method design, materials technology, statics and strengths of materials, and others, prepare specialists to manage automated manufacturing systems.

Job titles reported by graduates include manufacturing engineer, production control manager, engineering technician, applications instructor, and operations manager as well as other positions of responsibility in management and engineering-related areas. For more information, contact the School of Technology, Eastern Illinois University, Charleston, Illinois 61920.

Gateway Technical College. Robotics-related courses are part of two programs, each leading to the degree of associate of applied science. The newly emerging field of electromechanical technology and the associated robotics equipment are covered in electromechanical technology. Topics include manufacturing processes, principles of electrical and hydraulic systems, robotics mechanics, programmable controllers, and control systems.

Computer Integrated Manufacturing (CIM) Technician is a two-year associate degree program designed to prepare technicians for employment in manufacturing facilities. Students develop skills in computer operation, integration of computer numerical control (CNC) machine tools, computer aided design, and computer aided manufacturing. They also become competent in robotic applications and gripper design and in material handling techniques.

Graduates of the CIM manufacturing program have found entry-level employment as computer integrated manufacturing engineer/technicians, responsible for the integration of equipment involved in automated manufacturing operations; as manufacturing engineering technicians, helping to upgrade all manufacturing areas related to computers; as robotic specialists, programming and editing electromechanical devices; and as manufacturing technicians, using special knowledge of CNC machine tools. Major courses in the electromechanical technology (robotics) sequence are taught at Gateway Technical College Kenosha campus; major courses in the computer integrated manufacturing technician sequence are taught at the Racine campus. Selected courses may be taken at either Kenosha or Elkhorn campuses.

For additional information, write Gateway Technical College, Industry Division, 3520 30th Avenue, Kenosha, Wisconsin 53144-1690.

Henry Ford Community College. Students who choose the automation/robotics option of the electrical/electronics technology program earn an associate in science degree. The option emphasizes four key technical areas: electrical/electronic power and controls; hydraulic/pneumatic power and controls; computer circuitry and programming; and industrial instrumentation and electronic calibration. Graduates are prepared as support technicians in research and development, assembly and testing, field service, and product sales.

In the course Power Systems and Mechanical Interfaces to Automation and Robots, students explore the relationships between fluid power systems and mechanical devices driven by fluid power actuators. In Automation Controls and Robotics, control specifications written by the student for laboratory automation machines and industrial robots are used to implement controls for nonsynchronous and synchronous operation of the machine.

For information, write Henry Ford Community College, 5101 Evergreen, Dearborn, Michigan 48128-1495.

Illinois Central College. The robot and automated manufacturing technology program teaches the installation, programming, troubleshooting, and application of industrial robots in the manufacturing environment. Illinois Central College offers a two-year degree of associate in applied science. Courses combine lecture and hands-on training in a laboratory with four full-size industrial robots. Most graduates find employment as robotic technicians or manufacturing technicians who work with robots. For information, write Illinois Central College, East Peoria, Illinois 61635.

Indiana Vocational Technical College (popularly known, even to its administration, as Ivy Tech). This college offers an associate of applied science and an associate of science degree in automated manufacturing technology. Students in the AMT program take courses in introduction to robotics and work cell design. They learn about classification, programming, maintenance, and application of robots. The program emphasizes the application of robotics to manufacturing and interfacing the control systems with the robot's environment. Additional course work includes the integration of robots and automated systems within an advanced manufacturing system. The use of machine vision systems, bar code scanning systems, programmable controls, automated storage and retrieval systems, automation management, and software for computer integrated manufacturing comprise the rest of the program. Ivy Tech reports that 100 percent of its 1991 graduates in the automated manufacturing technology program found employment, with an average annual starting salary of $19,760. For information, write Division of New Technologies, Indiana Vocational Technical College, 7377 South Dixie Bee Road, Terre Haute, Indiana 47802.

Iowa State University. Robotics is a part of the manufacturing curriculum, which itself is under the industrial engineering program. A four-year technology program, the manufacturing program is designed to prepare graduates for the manufacturing industry. Technical content courses include studies in automated manufacturing, CAD, advanced processing, digital and microprocessor electronics, and statistical process control.

Mandatory internships or co-op experiences help students translate classroom and laboratory experiences into the reality of industry settings. The co-op program helps students obtain career-related jobs for a total of 12 to 16 months while they obtain their education. Students alternate classroom studies with supervised work experience.

For more information, write Department of Industrial and Manufacturing Systems Engineering, Iowa State University, 205 Engineering Annex, Ames, Iowa 50011.

Lima Technical College. Flexible manufacturing systems (FMS) is a major area of study under the mechanical engineering technology program. The FMS program was chosen as a Center of Excellence by the Ohio Board of Regents.

The flexible manufacturing systems major provides a mix of courses from both the mechanical and electronic technology disciplines, with an emphasis on hands-on as well as lecture class experience. Students gain knowledge in flexible manufacturing cells, robotics, computer-aided drafting, computer-aided manufacturing, automated storage and retrieval systems, automated guided vehicles, and programmable controllers.

The engineering technician with this two-year associate degree has a broad background and is capable of programming CNC machine tools, robots, programmable controllers, and computers. Skills such as troubleshooting, interfacing systems, and maintenance also are learned. Drafting and design (either mechanically or computer-aided), hydraulics, pneumatics, mathematics, and physics are also studied.

Each year, FMS students complete a major project. These have included designing and building robots for competition or display at local, state, and national events, such as the Society of Manufacturing Engineers robotics expositions and contests.

For additional information, write Dean, Engineering Technologies Division, Lima Technical College, 4240 Campus Drive, Lima, Ohio 45804.

Lorain County Community College. The automation engineering technology program includes course work in robotics, sensors, work cell interfacing, flexible manufacturing systems, and computer-integrated manufacturing (CIM). Initial instruction uses small training robots; full-size industrial robots are part of the

CIM laboratory. Advanced classes explore the feasibility of inter-connecting robotic systems to personal computers as well as to programmable logic controllers.

Students can earn their associate of applied science degree in automation engineering technology. Options available include majors as systems specialist or maintenance/repair. Those who choose to earn the degree electronic engineering technology can major in applied electronics or in computer systems.

For additional information, write Lorain County Community College, 1005 North Abbe Road, Elyria, Ohio 44035.

New Jersey Institute of Technology (NJIT). Undergraduate and graduate courses are offered in robotics, CAD/CAM, and manu-facturing studies. Noteworthy is the school's Consortium for CAD/CAM Robotics, founded in 1982 to serve as a center for applied advanced research and technology transfer in automated manufacturing. The consortium has one of the largest university installations of research robots in the United States.

Interdisciplinary in nature, the consortium includes NJIT faculty from the departments of mechanical and industrial engineering, electrical engineering, and computer and informa-tion science. Included in its mission: education and training on robotics, including seminars, lectures, conferences, and specially organized short courses for engineers and managers.

For further information, contact Consortium for CAD/CAM Robotics, New Jersey Institute of Technology, University Heights, Newark, New Jersey 07102.

North Iowa Area Community College. Students may study robotics courses within two programs: automated systems technology and electronics engineering technology. Each leads to an associate of applied science degree. Graduates may transfer to baccalaureate programs in such fields as electromechanical systems, engineer-ing technology, or supervision and management.

The automated systems technology program prepares students in the electromechanical skills necessary to install, maintain, program, troubleshoot, and service high technology systems found in computer-automated manufacturing facilities. Course work includes studies pertaining to material handling, forming, shaping, assembling, and testing in product or process industries. Graduates may find work as technicians in systems/maintenance, instrumentation, electromechanical systems, control systems, robotics, and computer automated process control.

Electronics engineering technology is an associate of applied science degree program designed to prepare graduates for immediate employment with manufacturers of electronic equipment and as electronic maintenance personnel in manufacturing settings. Graduates may find work as technicians in computer labs, electronics, industrial processes, and industrial maintenance and as electronics equipment repairers.

For more information, write North Iowa Area Community College, 500 College Drive, Mason City, Iowa 50401.

Northern Illinois University. Robotics courses are included as part of the manufacturing technology program. Northern's B.S. degree with a major in manufacturing technology combines theory with laboratory experience to provide a hands-on education. This degree provides instruction in drafting, machining, plastics, and welding while integrating computers into manufacturing. Most graduates go to work for small- to medium-sized manufacturers, where they introduce new technology and solve production problems. Courses related to robotics include: Automatic Identification (methods and systems used to identify objects automatically, such as bar coding, optical character recognition, voice data entry, and vision); Automated Manufacturing Systems; Programmable Electronic Controllers; and Engineering Automation. For additional information, contact Department of Manufacturing Technology, College of Engineering and Engineering Technology, Northern Illinois University, DeKalb, Illinois 60115-2854.

Northern Kentucky University. The department of technology offers associate and bachelor degree programs in industrial technology and bachelor degree programs in manufacturing engineering technology and electronic engineering technology. The industrial technology degree programs are structured toward preparing an individual to manage the processes of production. The engineering technology degree programs prepare individuals to help in the technical development of production. Both degree areas require courses in robotics as well as courses in automated integrated production (computer-aided design (CAD), computer-aided manufacturing (CAM), computer numerically controlled (CNC) tools).

Both degree areas are oriented toward an application approach and have heavy laboratory requirements. Students pursuing an engineering technology degree have a co-op requirement; industrial technology students are strongly advised to include the co-op experience.

For additional information, contact Northern Kentucky University, Department of Technology, Highland Heights, Kentucky 41076.

Northern Michigan University. Within the department of industrial technologies and electronics, this university offers associate degrees in electromechanical technology and electronics technology as well as baccalaureate degrees in electronics technology, industrial technology, and electronics engineering technology. These degrees provide the necessary concepts, theory, and hands-on experience for students interested in the robotics and automation fields to seek work in manufacturing industries that use these technologies.

Courses include robotics and automation systems, computer-aided design (CAD), computer-aided manufacturing (CAM), flexible manufacturing systems, manufacturing resource planning II, quality control, and industrial safety and ergonomics.

For additional information, contact College of Technology and Applied Sciences, Northern Michigan University, Marquette, Michigan 49855.

Rensselaer Polytechnic Institute. Rensselaer has a strong robotics and automation course. Research strengths are distributed among the Center for Manufacturing Productivity and Technology Transfer, the Center for Advanced Technology (in robotics and automation) sponsored by New York State, and the Center for Intelligent Robotic Systems for Space Exploration. Additional curricular and research strengths are available through the mechanical engineering, automated engineering, and mechanics department (MEAEM) as well as in the electrical, computer, and systems engineering department.

Major applications include manufacturing, dexterous manipulation, robotic aids for the handicapped, and automated assembly with space applications.

Within the Center for Intelligent Robotic Systems for Space Exploration (established by NASA), researchers study intelligent robotic systems—machines that can perform humanlike functions with or without human interaction. These systems are used for activities too hazardous for humans or too distant or complex for remote telemanipulators. The center coordinates the work of the robotics and automation laboratories.

Rensselaer's Center for Manufacturing Productivity and Technology Transfer focuses its research in several technology areas: advanced materials; processing, evaluation, and recovery; electronics manufacturing; enterprise integration; manufacturing control; and product/process design and support systems. For further information, contact Rensselaer Polytechnic Institute, School of Engineering, Troy, New York 12180-3590.

San Jose State University. Within the department of general engineering, students can design sequences of courses to meet a particular career objective, such as robotics. In-depth training can include a traditional departmental major, an interdisciplinary major that combines courses offered by various departments, or an innovative major, made up of a combination of courses not already proposed by the school. The innovative major must be designed with the help of an advisor, and its uniqueness should be based on a particular career goal selected by the student.

Ninety units in math, science, and engineering are the foundation of the program, which has as its objective the general fundamental study of engineering sciences as its broad base for the profession.

A robotics option, consisting of 15 units, includes courses in computer science analysis, computer organization, computer architecture, electromechanics, and theory of automatic controls.

For additional information, write General Engineering Department, School of Engineering, San Jose State University, One Washington Square, San Jose, California 95192-0080.

Southeastern Louisiana University. This university offers a bachelor of science degree in industrial technology, with an automated systems specialty. Courses include computer-aided manufacturing (CAM), a course designed to teach the use of the computer and peripheral equipment to create the data base for programming the operation of manufacturing equipment such as lathes, milling machines, drilling machines, and robots; and a course in industrial robotics, including the operation, installation, and maintenance of pneumatic robot systems, along with the operation and programming of the programmable controller.

For information, write Southeastern Louisiana University, Department of Industrial Technology, P.O. Box 847, Hammond, Louisiana 70402-0847.

Southern College of Technology. Known as Southern Tech, this college, a unit of the university system of Georgia, offers a four-year bachelor degree program in mechanical engineering technology. The school emphasizes hands-on laboratory experience in manufacturing processes and techniques, instrumentation and controls, and equipment and machinery performance testing and evaluation. Southern Tech also emphasizes meeting the needs of the industries prevalent in the Southeast.

Students take a common core of courses; those who choose may specialize in manufacturing technology. That program is concerned with manufacturing production processes and operations, tool and jig design, and the design and layout of manufacturing facilities.

A team of students competes in the Student Robotics/Automation contest sponsored by the Society of Manufacturing Engineers. In 1992, Southern Tech's entry took first place, defeating teams from more than 50 colleges and universities in the United States and Canada. The automated work cell they built allowed an IBM robot to place and then remove the 64 squares of a chessboard.

For information, write Mechanical Engineering Technology Department, Southern College of Technology, South Marietta Parkway, Marietta (Atlanta), Georgia 30060-3896.

Southern Illinois University-Carbondale. The department of technology offers accredited bachelor of science degrees in engineering technology and industrial technology and a master of science degree in manufacturing systems. The undergraduate programs offer a robotics course for seniors. The graduate program has a heavy emphasis on industrial robotics.

The engineering technology program is a laboratory-oriented program, with specialties in mechanical and electrical engineering technology. Technologists are prepared to work in the spectrum between the design engineer and the factory technician. The industrial technology major is designed to prepare management-oriented technical professionals in the economic enterprise system.

The College of Engineering has a cooperative education program available to all students.

For more information, write Department of Technology, College of Engineering, Southern Illinois University, Carbondale, Illinois 62901.

St. Cloud State University. The engineering technology program includes course work in robotics. Robotics I is a required course in the manufacturing program and a technical elective in the engineering technology-general emphasis, computer science major, industrial studies programs, and other majors.

The course is taught by lecture, with an extensive hands-on laboratory component. Lab exercises challenge students to design, set up, and modify existing equipment, instrument a robotic system, and perform an experiment using computer-controlled vision-assisted training robots.

Paid internships are available in the manufacturing and quality engineering technology area.

For more information, contact Department of Technology, Headley Hall 216, St. Cloud State University, St. Cloud, Minnesota 56301.

University of Florida. Centered in the mechanical engineering department, the Center for Intelligent Machines and Robotics (CIMAR) is an interdisciplinary research group of faculty and students from the departments of mechanical engineering, electrical engineering, agricultural engineering, nuclear engineering sciences, and computer and information sciences. The center's specific research strengths are in the following areas: three-dimensional geometry and kinematic analysis of robotic systems, screw theory as applied to position and force control of robot manipulators, real-time computer graphics simulation, integration of computer systems (telepresence system development), and the control of remote robotic systems in hazardous environments. Funding for projects comes from a variety of sources, including the National Science Foundation, the U.S. Air Force, and the U.S. Department of Energy.

The University also conducts research in manufacturing at its machine tool laboratory. Projects there involve testing and evaluation of sensors for cutting performance of machining centers.

For additional information, write Center for Intelligent Machines and Robotics, 300 MEB, University of Florida, Gainesville, Florida 32611-2050.

University of Massachusetts at Amherst. The mission of the laboratory at this university is to conduct research on automatic robot programming and the use of design data in robotic process plan generation. Approximately 15 research associates and graduate students per year work primarily on contract research for general industry. Major areas of research include robot task-level language (50 percent), process planning for robot programming (30 percent), robot design (10 percent), and vision research (10 percent). Application areas include assembly, machine loading, material handling, and robot programming/task-level language development. For information, write Automation and Robotics Laboratory, Department of Industrial Engineering and Operations Research, 114 Marston Hall, Amherst, Massachusetts 01003.

University of Missouri-Rolla. Robotics and computer integrated manufacturing are taught here. The B.S. degree in engineering management constitutes the foundation for the computer-integrated manufacturing at this school. It is based on a three-plus-one formula, where three years' worth of technical manufacturing engineering course content is blended with a year's worth technomanagerial course content to synthesize an effective education in computer-integrated manufacturing.

The Flexible Assembly Cell features an assembly robot, various sensors, fixtures, and a feeder. The Flexible Machining and Assembly Cell features a CNC mill, a CNC lathe, a loop conveyor, a material handling robot, a cylindrical pick-and-pace robot, a quick changer for end effectors, an automated storage and retrieval system, and other components of automation.

For additional information, write Engineering Management Department, 223 Engineering Management Building, University of Missouri-Rolla, Rolla, Missouri 65401.

Wayne State University. Within the Division of Engineering Technology, the bachelor of science degree and the master of science degree are offered. The upper-divisional program admits students with an associate degree in an appropriate engineering technology discipline, an associate degree in engineering science, or color-level course work equivalent to an associate degree in an engineering/technology related area.

Major fields of emphasis are electrical/electronic engineering technology; electromechanical engineering technology; manufacturing/industrial engineering technology; mechanical engineering technology; and product design engineering.

At the graduate level, a course in robotics and flexible manufacturing is offered. Included: instruction in kinematics, dynamics, and control of manipulators, design and applications in flexible manufacturing cells, and computer-integrated manufacturing.

For more information, write Division of Engineering Technology, Wayne State University, Detroit, Michigan 48202.

A number of other schools offer courses in robotics and related technologies. Among them: Brown University, Carnegie-Mellon University, Indiana State University, Trident Technical College (Charleston, South Carolina), and University of Michigan. You will want to write schools that sound interesting, asking for additional information and visiting campuses wherever possible.

CHAPTER 9

JOB-HUNTING TIPS

If you want to work in robotics or related technologies, how do you get your first job in the industry? And, if you already have experience, how can you change jobs profitably? What can you do to improve your chances?

INDUSTRY PREDICTIONS

Although the United States has traditionally been a leader in manufacturing, the move toward a global economy has implications you can't afford to ignore as you plan your career strategy. For instance, Robert P. Collins, president and CEO of GEFanuc Automation North America, points out that Japan has 176,000 industrial robots while the United States has approximately 33,000. In 1992, a recent survey ranked the United States twentieth among industrialized nations in per capita consumption of numerically controlled machine tools. Between 1978 and 1989, U.S. exports of finished goods stagnated, Collins says, while imports of finished goods jumped from 66 percent to 77 percent.

While Collins calls for more government investment in advanced production technologies—technologies that presumably include robotics—he also predicts a downsizing of industrial organizations. "Industries must also become lean, quality-driven, fast-moving organizations focused on the customer," he suggests.

Does Collins's prediction mean fewer jobs for entering graduates? Maybe. But the U.S. government doesn't necessarily think so. In 1992, the secretaries of labor and commerce signed an agreement designed to help companies adopt new production methods. The Labor Department's office of work-based learning was developing a pilot training program that community colleges and others can use to train workers who are new to automated manufacturing technologies.

Other government efforts, begun in 1990, channeled manufacturing outreach efforts through small business development centers in six states. Maryland, Wisconsin, Pennsylvania, Texas, Missouri, and Oregon matched government funds with state moneys to underwrite small company use of manufacturing data bases.

No one can predict with certainty what the job market will be like when you are ready to start looking. But as long as you are realistic about your opportunities, as long as you are aggressively pro-active, as long as you work as hard at selling yourself as you have done in acquiring your skills, you have maximized your chances.

As Indiana Vocational Technical College (Ivy Tech) puts it, "The reason for getting a high quality technical education is simple—to get a job." Its graduate follow-up survey indicated that 100 percent of graduates in the automated manufacturing area had been placed, at an average starting salary of $19,760.

GET INVOLVED EARLY

Robotics veterans have several tips for people who want to enter the industry:

- Do your own homework. Read the trade press. You will find a selection of trade magazines listed in appendix B of this book. Read the business sections of your local daily newspaper. Keep up with developments in industry and technology—and their bottom-line implications for business—by reading *Business Week, Fortune,* and *The Wall Street Journal.* Don't expect your school counselor to know what is going on. Take that responsibility yourself.

• Become involved. Join trade associations, especially those with student chapters. Noteworthy are Robotics International, Computer and Automated Manufacturing (CASA), and Machine Vision Association of America, all divisions of the Society of Manufacturing Engineers (SME). The Institute of Industrial Engineers (IIE) is another association you will want to check out.

Both SME and IIE have professional chapters in major cities. Attending meetings and programs these organizations sponsor will give you not only a briefing on technology developments but also an opportunity for professional contacts.

Go to the trade shows, symposiums, and conferences—not only in robotics but in related fields, like CAD/CAM, machine vision, and other areas of manufacturing technology. You will find calendars of upcoming events in the trade magazines. For instance, *Managing Automation,* a magazine that covers computer integrated manufacturing heavily, listed 23 such shows in a single issue. Some were regional: for example, Northern California Plant Engineering and Maintenance Show and Material Handling and Packaging Show and Motion Control Technology Conference and Exhibition/West; others were national (National Robot Safety Conference and Autofact Conference and Exposition are examples.). Students with appropriate identification who are preregistered can usually attend for substantially less than regular fees.

At the shows, plan to spend time and effort talking to exhibitors, acquiring and studying product literature, and building your own information files. In addition, shows sponsored by associations or other professional groups often have job placement bulletin boards, listing opportunities available to members. Most of these positions require prior previous experience. But even though you may not yet have the skills you need, you will get an idea of just what is involved in these positions; you may want to broaden your education or experience to round out your qualifications.

- Broaden your sights. "You can't become so specialized that you become obsolete," warns Rebecca Stevens, editor of *Sensors,* one of the trade publications. Don't just think "and robotics"; think "flexible automation." Don't just think "machine vision"; think "electronic imaging."

  Up to now, you may not have considered warehousing as being related to robotics. Yet the International Conference on Automation in Warehousing, sponsored by the Institute of Industrial Engineers and designed for persons involved in handling, storing, and controlling products and materials, discusses topics like high performance picking. Pick-and-place robots are one of the technologies involved.

- Know the players, the companies, and the trends. Pick up the phone occasionally, calling someone whose name you know from trade press or conferences and asking for advice—a technique that works wells, if not overdone.

EMPHASIZE VERSATILITY

If you wait to start job hunting until just before you graduate from college, you are extremely late. Instead, from day 1, you can work with your placement office and with your departmental advisor to help make yourself employable. Take advantage of any opportunity you get to intern, regardless of whether the internship is paid. If your school has a co-op program, so much the better. Industry experience—along with a good recommendation on your performance—can help.

You will need to convince potential employers you know more than just robots or robotic technology. Instead, emphasize that you are interested in a manufacturing career. In your resume and cover letter, if you are a brand-new graduate, point out courses you have taken in related technology: hydraulics, pneumatics, CAD/CAM, and control systems. Math and language skills also are important. If you have previous experience, however, employment agencies advise you to skip the course listings in favor of on-the-job information.

FINDING A JOB

Finding your first job won't be easy or automatic. You will need to put considerable time and effort into making contacts and looking for potential employers. *Dun and Bradstreet, College Placement Council (CPC) Annual*, and similar publications have names, addresses, and profiles of companies that deal with high technology. Study them intensively. The profiles not only describe the companies in general terms but also often indicate specific areas to which you may want to apply. You will learn, for instance, that McDonnell Douglas Corporation has a number of opportunities: candidates with bachelor's, master's, or doctor's degrees in manufacturing engineering are encouraged to apply. At McDonnell Douglas, manufacturing engineers work on applications involving N/C machine controls, robotics, process technology, CAM, tool design, equipment evaluation, facilities, test equipment design, and production planning related to fabrication and assembly.

Chambers of commerce in technology-rich locations such as Orlando (an area in which many electronics, lasers, and related industries are located) often sell directories listing company names and addresses. For instance, you can write to the Orlando Chamber of Commerce, 75 East Ivanhoe Boulevard, P.O. Box 1234, Orlando, Florida 32802 for its *Directory of High Technology Companies*—a directory that is periodically updated.

Another good place to start is with the *Robotics and Vision Supplier Directory,* which can be purchased from Robotic Industries Association, 900 Victors Way, P.O. Box 3724, Ann Arbor, Michigan 48106. The directory is jointly published by Robotic Industries Association, Automated Imaging Association, and National Service Robot Association. It lists robot manufacturers and distributors, suppliers of accessory equipment, and vendors of systems integrators. Also included is information on machine vision: manufacturers, distributors, and suppliers of accessory equipment.

Employment Agencies

Once you have two or three years' experience in factory automation, getting a job related to robotics or switching companies still isn't easy. The job market in the nineties is tight; companies may not be hiring or (understandably) may be promoting from within. Union agreements spelling out seniority may cover who stays and who goes if layoffs are announced.

Major companies looking for certain training, skills, and experience often use employment agencies. Such agencies frequently advertise in the trade press. Fees for placement come from the employer—not from the job candidate. Many states regulate fees and conditions under which agencies doing business in the state must operate.

Most agencies prefer to work with candidates who have from two to five years' experience, and who hold four-year degrees. "Typically, I work with the B.S.M.E. (bachelor of science in mechanical engineering) for all mechanical aspects and the B.S.E.E. (bachelor of science in electrical engineering) for the electronics," says Jim Lakatos, president of IBA Personnel, a Michigan-based employment agency. "All four-year degrees don't carry the same weight with employers, though. I may have a candidate with a 3.9 grade point average on a 4.0 scale who has a four-year B.S. degree in engineering technology. Such candidates are difficult to place.

"A lot of employers don't like technology degrees. They feel that candidates took the courses because they were easier or required less math. Now that's not necessarily so. If you're the best designer in a company, it doesn't matter what your degree was. However, if you have a technology degree, it may cost you opportunities when you try to switch companies."

Lakatos doesn't handle first-time-out candidates. Primarily, he places candidates with up to five years' experience at salaries ranging from $30,000 to $42,000. Too much experience with one firm can be a turnoff to potential employers, he says. Lakatos tells of hearing a well-known national recruiter speaking at a trade seminar say he won't deal with anyone who has worked more than 6 years with one company. "It's difficult for such a person to leave, to be flexible enough to move to a new employer," the recruiter argued.

One of the most important things for candidates to remember is that their interests and the interests of the employment agency are not necessarily the same. Agencies find people for jobs; they do not find jobs for people. Lakatos reminds candidates that he is looking for a person to fill a corporate vacancy. "The company is paying me," he says. "The candidate isn't."

Nationwide Networking

For a candidate with experience, one of the most important reasons for signing on with an agency is the working agreement the agency may have with others in the field. IBA Personnel, the agency Lakatos runs, belongs to a network called Intercity Personnel, based in Wisconsin. "Although the network has professional level openings," he says, "over 90 percent of the placement orders we receive are for engineering candidates.

"If you send me your resume and you're one of my top candidates," he says, "I'll fax it out so that roughly 240 other agencies in my network have your resume." Intercity Personnel is perhaps the third largest network in the United States, Lakatos explains. Another network has over 400 agencies.

IMPROVING YOUR RESUME

The resume you send an agency or a prospective employer should be the strongest possible. Of course it should look its best, no matter how much time and effort it takes. But you don't have to spend a fortune. Lakatos says having your resume typeset or printed up on fancy paper is a waste of money. White paper and good-quality typing or good-quality printout is all you need, he explains. Don't use a dot-matrix printer if you have alternatives; the quality of the letters and type not only doesn't measure up to employers' expectations but also will not fax well. If you must use a dot-matrix printer, change the ribbon frequently. Even still, typing is better than dot-matrix, Lakatos believes.

Here are some do's and don'ts:

- Make your resume visually attractive.

   If you were an employer seeing the resume for the first time, would you want to read it? Could you? "I've seen candidates so hung up on getting their resumes on one page that they have $^1/_4$ inch margins, scrunched-together type, and almost no white space," Lakatos says. "I've heard that when a resume hits an employer's desk, it's typically looked at for seventeen seconds the first time around. If it's hard to read or looks cluttered, the resume is usually trashed. You want your resume in the pile put aside to be read later."

   Don't go to the other extreme, though. Some candidates send Lakatos resumes more than a dozen pages long; each college gets a separate page, as does each employer. You also don't need to list the courses you have taken.

   The ideal length? If you are a just-out-of-school, no-experience candidate, one page. If you have experience, two pages.

   And don't use dark-colored paper. It won't fax well. Hold off on the fancy script or elite type, but bold or heavier type is okay to use for key words.
- Do include dates.

   Employers want to know when you graduated from college and when you worked for particular companies. "If you don't include dates," Lakatos says, "prospective employers will get the impression you're over 55 with 20 or more years' experience at the same company, probably not flexible enough to move on."

   It is okay to say you have gaps in the dates because you have been laid off or dropped in a company reorganization. Just put that information up front in your cover letter, Lakatos advises. Sometimes you may be hired for a company, work only six or seven months, and then be let go. Include the dates, nevertheless.

"If you indicate in your cover letter that you were hired at the time the company was making reductions," Lakatos says, "most employers will recognize that you were hired to solve specific problems—you could provide expertise that they didn't have in-house—and that after you'd solved the problems, you were dropped."

- Include a list with a few of the company's major products when you mention former employers.

Because so many organizations have merged or are part of multinational corporations, a prospective employer may not immediately recognize the names of companies you have worked for. If you say you worked for XYZ Corporation, manufacturer of hydraulic pumps, an employment agency personnel department can tell right away if you have the type of experience that might match their needs. Otherwise, they will have to check an industrial directory to see what the company makes.

Also, if you worked on a particular product, say so—in detail. Were you a machine designer, a control designer, or a project engineer? If you put down that you were a project leader or project manager, indicate what type of project it was that you headed.

- Use verbs. Lots of them. Be as specific as possible.

Don't say you participated in the development of a product, even if you did very little as a team member. Instead, say, "I, along with three others, designed a widget." "I designed the simple mechanical aspects of the robotic system." Don't say, "I was involved in . . ." Instead, tell exactly what you did.

- In your cover letter, indicate that you have transcripts available, references from companies (name them), and a detailed list of projects you have worked on. Don't include them with the cover letter and resume that go to the employment agency, Lakatos advises. Instead, say you will be glad to send them on if they are wanted. The agency will often request these credentials and will use them to see how you match up against what its clients (the hiring companies) have in mind.

When you are sending your resume to an employment agency, Lakatos advises, ask what network it belongs to. Don't blindly send out a hundred resumes to agencies. Many of them may belong to the same network, and you are just duplicating efforts. "It turns me off a candidate if I find that 33 of his or her resumes are already in my network," says Lakatos.

If you are not willing to relocate, use a local employment agency. Call for an appointment; take along samples of things you have done: CAD drawings, pictures of machines you have designed, or even a small model. If you are going to look for work on a national basis, contact the employment agencies. Ask up front if they have people who are working in the geographical locations where you want to be.

PHONE CONTACT

For candidates who have contacted an agency like Lakatos's by phone, mail, or fax, an important screening comes when an employment counselor calls back. That initial phone contact can make a big difference in the agency's interest.

Most placement counselors do a fast interview over the phone to find out if a candidate whose resume and cover letter sound promising actually meets the guidelines a company has given. A few key questions—and the candidate's response—may determine whether the agency is interested in making a match. Says one counselor, "I'll ask you why you want to switch jobs. You'd better tell me you want to make a career move for advancement or get a better job or go to a different company in order to further your career.

"If you tell me you want to change jobs to make more money, that's a turnoff. If you're a whiner—if you tell me you don't like your boss—that's a no-no. If you're dissatisfied with your job, be dissatisfied for a concrete reason. Maybe you've reached the limit in what you can do in your present company. That's valid."

Legitimate reasons for job-switching that agencies will usually accept are company mergers or internal consolidation. "I know a company that closed its plant in California but kept its Michigan operation open," says Lakatos. "In some cases, the jobs were redundant. They didn't need two purchasing managers for the Michigan plant. But the California location had been making a product component, and the Michigan site had to take over. The company needed hydraulic pump designers and project engineering managers in Michigan, but it couldn't just transfer everybody. Some key people refused to leave California. They eventually found new jobs out there, and the company hired appropriate personnel for the Michigan location."

YOUR INTERVIEW WITH THE COMPANY

Occasionally, outstanding candidates may be invited directly to the company's factory or offices for an initial on-site interview. However, the usual next step in screening is a phone interview at a prearranged time with a representative from the manufacturer. The employment agency sets up the interview.

Whenever possible, Lakatos likes to brief his candidates before the initial phone interview. He will fax the candidate the job description and, when time permits, complete background on the company. Usually the company's initial representative who makes the phone call is a staffer from the human resources department. If candidates still look promising, their names are passed on to the engineering department, who follows up—either by phone, by an invitation to visit the company, or both. "Don't gripe about your former employer," Lakatos says. "Concentrate on presenting your experience. Communication skills help. No one expects you to be a hotshot at sales, but you do have to be able to tell an interviewer exactly what you are doing."

Even if you get the company background from the agency, do your homework. Candidates should be able to discuss companies intelligently. That means library research to find out what the company makes, where it stands in the marketplace, and its organization and financial status. A good place to locate the information is in the company's annual report. You may also want to check *The Wall Street Journal* index or the *New York Times* index to find out if the company has been mentioned recently.

An interview is a dialogue. You have to show you are really interested. A candidate has to hustle and fight and work hard to get the job. You will need to ask questions that show you don't want to be stagnant in your career growth.

"At the end of the interview, or if you're asked what questions you have, show the interviewer that you're interested in improving yourself," advises Lakatos. "Ask if the company has a tuition reimbursement plan. Ask about seminars, conferences, in-service training. Tell the interviewer you want to take courses and gain knowledge that will make you more valuable for the company. In an interview with a company that may hire you, don't ask about retirement benefits or number of weeks of vacation. You can get that information from the human resources department—after you're offered the job."

Even though you want to get into robotics and can certainly mention your interest in the technology, don't get hung up during the interview on specific job tasks. "People who say, 'I'll do anything you need; I want to work for your company' get hired a lot quicker than persons who tell the interviewer about what they want to do four years from now," Lakatos warns.

Do write a good, positive follow-up letter, and mail it promptly. If you have met several people on-site, write each individually. Counselors say they have known several candidates who have received job offers because of the quality of their follow-up letters.

THE LONG-TERM PICTURE

The job market is tough, especially as companies downsize in response to a sluggish economy. "Companies are looking," says Lakatos, "but they're doing more shopping than buying. On the other hand, people are still retiring, finding other jobs, going through divorces and wanting to leave town, and—for whatever reason—resigning. Most organizations have holes to fill. Agencies like ours are always looking for one person, here or there. That person may be you."

CHAPTER 10

WOMEN AND MINORITIES
IN ROBOTICS

Women and members of minorities who want a career in robotics or related technologies certainly have a chance—if they are qualified.

That is a big "if." Companies are willing to hire and promote women and minorities if they have the training that is required to perform. By the year 2000, according to Bureau of Labor Statistics (BLS) predictions, more than 80 percent of the labor force growth will be attributed to women, minorities, and immigrants. The labor pool will be aging, and companies will require higher levels of analytical skills.

BLS and others predict that by the year 2010, the United States may experience a shortfall of over 550,000 engineering and technical professionals. No one knows exactly how many women and minority engineers and technicians are working with robots or with related technologies. Since many robotic industries personnel have engineering backgrounds, however, it is useful to look at the employment record for engineers.

OPPORTUNITIES FOR WOMEN

In the past, women interested in engineering often had little or no encouragement to develop manual skills and were unused to handling tools or technical equipment. One important reason some women have not gone into science and related fields may be their perception of how much time they will spend in the labor force. If a young woman expects to spend a substantial amount of time out of the work force because she is rearing children, she has less incentive to choose career options that require substantial educational commitment.

Today, though, most young people assume that women will combine a job outside the home with family-rearing responsibilities. However, many young women do not realize that substantial numbers of higher-paying jobs require a more quantitative educational background. They pass up science and math courses early in life, thus inadvertently closing the doors on many financially rewarding and enjoyable jobs they might like to hold.

EARLY CHOICES

Based on a Rockefeller Foundation study by Sue E. Berryman entitled "Who Will Do Science?", the Committee on the Status of Women of the American Physical Society has reported some disturbing conclusions. "By ninth grade," it says, "over one-third of those who will later earn a quantitative bachelor's degree already expect to pursue a career in science. By the end of twelfth grade, the pool is fully established. A necessary component of the pool's educational profile is the completion of advanced high school math. This optional sequence, which is elected by one-third less girls than boys, marks the first point at which the educational profiles of the two sexes diverge."

Young women, then, are skipping courses that could lead them to well-paying and challenging science careers—including careers in robotics and related fields. In particular, they are turning away early from the math they need to succeed.

Two Critical Decision Points for Women

The Committee on the Status of Women in Physics (CSWP) has reported two major decision points for women—points where choices must be made about future educational investment.

The first decision point comes much earlier than most students realize. By ninth grade, over one-third of those who will later earn a bachelor's degree in quantitative fields (math, physical or biological sciences, computer science, engineering, and economics) already expect to be scientists. By the end of twelfth grade, the pool is fully established—that is, essentially all those who will go on to a bachelor's degree in quantitative science have already decided to do so.

To succeed in any of these technical fields and to work in robotics, you need a strong math background. Yet advanced high school math is elected by one-third fewer females than males.

Of those who completed the necessary high school math sequence, according to the Scientific Manpower Commission, only 21 percent of the young women, as compared to 51 percent of the young men, chose a quantitative field when declaring their undergraduate majors.

These two turning points account for more than two-thirds of the women eliminating themselves from quantitative fields compared to their male peers all the way through the Ph.D. degree.

The AAUW Study

A highly publicized study, released in 1991, adds further information about why high school girls may tend to drop math courses earlier in their studies than boys do. The study, commissioned by the American Association of University Women (AAUW), was designed to examine the interaction of self-esteem and education and career aspirations in adolescent girls and boys.

The survey of 3,000 children between grades four and ten in 12 locations nationwide focused on the differences in attitudes between girls' and boys' perceptions of themselves and their futures, measured the changes in attitudes as adolescents grew older, and identified critical processes at work in forming adolescents' attitudes of self-esteem and identity. It looked at adolescents' career choices and expectations and their perceptions of gender roles and at the part the educational setting plays in that. Finally, the survey examined the relationship of math and science skills to the self-esteem and career goals of the boys and girls in the study.

One of its most significant findings: how students come to regard math and science differs by gender. "Math and science have the strongest relationship on self-esteem for young women," the study says, "and as they 'learn' that they are not good at these topics, their sense of self-worth and aspirations for themselves deteriorate."

The survey found that at all grade levels, adolescent boys are much more confident than young girls about their abilities in math. By high school, one in four males—but only one in seven females—say they are good in math. The study concludes that girls interpret their problems with math as personal failures while boys project their problems with math more as a problem with the subject matter itself.

Another significant finding of the AAUW study was that students who like math and science are more likely to want careers as professionals. They are more likely to name professional occupations as their first career choice.

A second AAUW report, "How Schools Shortchange Girls," researched by the Wellesley College Center for Research on Women, is a study of major findings on girls and education. "A well-educated work force is essential to the country's economic development," the report says. "Yet girls are systematically discouraged from courses of study essential to their future employability and economic well-being." By the year 2000, the U.S. work force will require strengths and skills in science, mathematics, and technology—subjects, the report says, that girls are still being told are not suitable for them.

Among the report's recommendations: "Girls must be educated and encouraged to understand that mathematics and the sciences are important and relevant to their lives. Girls must be actively supported in pursuing education and employment in these areas."

Copies of these and other AAUW publications are available for purchase from the American Association of University Women, Sales Office, P.O. Box 251, 9050 Junction Drive, Annapolis Junction, Maryland 20701-0251.

SOCIETY OF WOMEN ENGINEERS

For more than 40 years, the Society of Women Engineers (SWE) has actively encouraged women to pursue and excel in engineering careers. The number of women in engineering, although still small, has increased over the years to the point that the society has over 5,000 professional members and 10,000 student members.

However, since 1987, fewer women have been entering engineering degree programs. In 1992, women made up only 15 percent of the engineers who entered the work force and only 4 percent of all practicing engineers. The figures are even lower for minorities. As a result, SWE is actively developing outreach programs to encourage minorities to enter science and engineering. Yet in 1989, one-fifth of all the minority engineering bachelor's degrees granted in the 50 states and the District of Columbia were produced by just ten institutions. That year, African Americans and Hispanics each made up less than 5 percent of those earning bachelor's degrees in engineering; Asian Americans made up just under 10 percent.

The SWE Scholarship Program

As part of its national educational activities, SWE administers approximately 40 scholarships, totalling more than $60,000 and varying in amount from $1,000 to $4,000. All SWE-administered scholarships are open only to women majoring in engineering in a school, college, or university with an accredited engineering program. They must be in a specified year of study during the academic year after the grant has been presented.

SWE has two national scholarship programs. Applications for freshman scholarships and reentry scholarships are available from March through May. Completed applications, including all supporting materials, must be postmarked no later than May 15.

Freshman scholarships include:

- The Westinghouse Bertha Lamme Scholarships, established in 1973. Recipients must be U.S. citizens.
- The General Electric Foundation Scholarships, first awarded in 1975. Scholarships can be renewed for three years if students show continued academic achievement. In addition, the General Electric Foundation provides $500 for each entering freshman recipient so that she can attend the annual National Convention/Student Conference and provide support to her local section. Recipients must be U.S. citizens.

- The Admiral Grace Murray Hopper Scholarship, established in 1992, for female freshmen entering the study of engineering or computer science.
- The Anne Maureen Whitney Barrow Memorial Scholarship, for an undergraduate woman entering engineering or engineering technology—renewable yearly until completion of the undergraduate degree.
- The TRW Scholarships, established in 1974. Scholarship winners are chosen by the best national, regional, and new student sections.

Another scholarship has been established to recognize the fact that women engineers who have interrupted their careers—perhaps for marriage or family commitments—can have special needs. The Olive Lynn Salembier scholarship, honoring a previous SWE president, helps women who have been out of the engineering job market a minimum of two years to get the credentials necessary to re-enter the job market as an engineer.

SWE also has a number of scholarships for sophomore, junior, senior, and graduate engineering students. Applications for these scholarships are available from October through January only. Completed applications, including all supporting materials, must be postmarked no later than February 1.

Some of these scholarships require special qualifications. For instance, the Digital Equipment Corporation Scholarship, awarded at the end of the freshman year, goes to a woman student member majoring in electrical, mechanical, or computer engineering who is attending a university in New York or New England. The Stone and Webster scholarships, for junior or senior female engineering students, are awarded (one each) to students in chemical, civil, electrical, environmental, and mechanical engineering.

The Judith Resnik Memorial Scholarship (established in honor of SWE member Judith Resnik, the astronaut who was killed in the explosion of the *Challenger* space shuttle) is awarded to a senior and SWE member studying an engineering field with a space-related major who expects to have a career in the space industry.

The General Motors Foundation scholarships, established in 1991, are awarded to students entering their junior year and majoring in one of the following: mechanical, electrical, chemical, industrial, materials, automotive, or manufacturing engineering or engineering technology. Recipients must have a minimum cumulative grade point average of 3.2 on a 4.0 scale, demonstrate leadership characteristics by holding a position of responsibility in a student organization, and exhibit career interest in the automotive industry and/or manufacturing environment. The scholarship is renewable for the senior year.

The General Motors Foundation also funds a scholarship with similar requirements for a first-year masters-level student.

Past winners of scholarships reflect diverse backgrounds and interests. For instance, J. Cameron Crump, of Spokane, Washington, has been interested in engineering as a career since junior high school, taking such hands-on courses as auto mechanics, metal shop, and welding in addition to being cocaptain and manager of her high school tennis team and president of the drama club. Cameron, who is majoring in structural engineering, plans to take four years of Chinese language studies also. After graduation, she hopes to join a company that can use her engineering skills on international projects.

Jennifer Martin of Tulsa, Oklahoma, an aerospace engineering student who has her advanced ground instructor license, helped found the Notre Dame Flying Club. Seema Jayachandran of Salinas, California, a junior in electrical engineering at the Massachusetts Institute of Technology, has worked summers as a computer programmer at the U.S. Naval Oceanographic and Atmospheric Research Laboratories. Heather Bell designed electrical instrumentation, conduit routing, and control circuits at the ALCOA Ingot Plant while still a student in computer and electrical engineering at Purdue University.

Winner of the Olive Lynn Salembier reentry Scholarship Cynthia S. Kustin, of Anchorage, Alaska, returned to school after a 20-year absence. During her first three years of engineering school, she attended part-time, commuting 150 miles round trip.

Deadline Dates

Deadline dates for applying for these scholarships are extremely important, warns SWE. Applications postmarked after the submission dates are not considered. In addition, because many applications are received, SWE notifies scholarship recipients only. Those receiving freshman and reentry scholarships are notified by September 15; those receiving sophomore, junior, senior, and graduate scholarships are notified by May 1.

For further information and to receive scholarship applications, send a self-addressed, stamped envelope to the Society of Women Engineers, United Engineering Center, Room 305, 345 East 47th Street, New York, New York 10017. As part of the application, you will be required to submit transcripts; letters of reference; and an essay on the reasons you have decided to study engineering, why you have chosen your major, and why you are applying for a scholarship.

MINORITY OPPORTUNITIES

The Engineering Manpower Commission of the American Association of Engineering Societies (a multidisciplinary organization of engineering societies) evaluates trends in the engineering work force. The commission tracks how many women and ethnic minorities are participating in engineering.

In 1950, when over 50,000 bachelor's degrees in engineering were awarded by U.S. colleges and universities, most students were men; only 0.2 percent of the B.S. degrees in engineering went to women. However, 40 years later, 15.3 percent of the B.S. degrees in engineering were earned by women. African Americans, Hispanic Americans, and Native Americans all have gained; in fact, the share of all bachelor's degrees in engineering awarded to these three ethnic groups has doubled since the early 1970s. Yet the numbers—especially for American women of color—are small. In 1991, 70 percent of all women earning bachelor's degrees in engineering were white, nonminority Americans, and only 2,505 B.S. degrees were awarded to American women of color. The commission also tracks Asian Americans, but they are not underrepresented in engineering.

As part of the Society of Women Engineers's strategic plan for 1993–1995, the organization has made a commitment to develop outreach programs that will encourage underrepresented minorities in science and engineering.

A Higher Education Outreach program, piloted in 1989, sponsored by SWE in cooperation with its local sections and funded by a grant from NASA, had three components:

- A one-day experience that combined a university-based engineering project with an industrial tour. For instance, SWE's Central New Mexico section featured robotics as its theme for an apprentice day. Twenty-five minority girls from six high schools toured Sandia National Lab's robotics facility and gained hands-on experience at the University of New Mexico's robotics work stations.
- A one-week summer residential program at a university that exposed students to four engineering disciplines and introduced students to college living

- A mentoring approach that linked women engineers and women engineering students to junior and senior high school minority women with an aptitude for math and science

The overall goal of the program was to increase the number of black, Hispanic, Native American, and Pacific Islander women in engineering careers.

MINORITY RECRUITMENT

Many engineering schools and companies recruit hard for qualified candidates. Sometimes the recruitment even starts before high school. A nationwide mathematics coaching and competition program, aimed at combating math illiteracy, is open to seventh and eighth graders; for information, contact the National Society of Professional Engineers (NSPE), Information Center, 1420 King Street, Alexandria, Virginia 22314.

Most engineering colleges and universities have special programs for culturally disadvantaged students. Minority students often take advantage of these programs. One program which has received nationwide recognition is Georgia Tech's freshman engineering workshop, which invites ninth-grade students from minority backgrounds to visit the campus, meet with faculty, visit companies that employ engineers, and plan careers in engineering or related fields. For additional information on this program, write Director of Special Programs, College of Engineering, Georgia Institute of Technology, Atlanta, Georgia 30332.

Similar enrichment, awareness, or recruitment programs often exist at universities in metropolitan areas. Illinois Institute of Technology has summer programs for talented, minority, Chicago-area high school juniors interested in math, science, or engineering. Some scholarships are available. Write to the Illinois

Institute of Technology, Engineering Department, 3300 South Federal Avenue, Chicago, Illinois 60616.

At the University of Texas-Austin, the Equal Opportunities in Engineering program includes scholarships. Write to the University of Texas-Austin, Equal Opportunities in Engineering, College of Engineering, ECJ2-102, Austin, Texas 78712.

Sometimes corporate scholarships are set up for students from certain ethnic backgrounds. For instance, in 1991, Amoco Company established the Amoco/MACSTA Youth Award. The cash award is given to a Chinese American youth who is a high school junior and is interested in majoring in the field of science and technology. Selection is based on the applicant's academic performance, honors and awards received at school, and his or her college major. Deadline for applications is mid-September.

For additional information, write Committee Chair, Amoco/ MACSTA Youth Award, 2565 Riverwoods Road, Riverwoods, Illinois 60015.

To locate such programs, ask your school guidance counselor for details, or write to an affirmative action officer at a college or university near you.

Another excellent source of information is *MEPS USA: The Directory of Precollege and University Minority Engineering Programs.* Available for purchase and inexpensive, the directory covers more than 50 precollege and university engineering programs for minority students, geographically arranged. It is published in July of even years. For information, write National Action Council for Minorities in Engineering, Three West 35th Street, New York, New York 10001.

SPECIAL HELP

A number of organizations provide help, guidance, materials, and even scholarships to women and minority students who want to be engineers. While they aren't necessarily focused on robotics and robotic technology, nevertheless they are worth looking into.

Films about engineering your teacher can order directly and show in class include:

- *Bridging New Worlds (Uniendo Nuevos Mundos),* a film in English with Spanish subtitles, helps parents and encourages their children to study engineering and the applied sciences. For information, contact Bilingual Cine Television, 2601 Mission Street, Suite 703, San Francisco, California 94110.
- *Technology Occupations* (order #VT-02) is a videotape for junior and senior high students that highlights careers in drafting, plastics, computer sciences, electrical technology, programming, and tool building.
- *Engineering Disciplines* (order #VT-01), also for junior and senior high students, highlights careers in structural, electrical, manufacturing, aeronautical, mechanical, and chemical engineering. Information on ordering these videos is available from JETS, the Junior Engineering Technical Society, 1420 King Street, Suite 405, Alexandria, Virginia 22314-2715. JETS also has career literature available.

JETS can also give you information on the National Engineering Aptitude Search. While the exam is not specifically aimed at women and minorities, it is a guidance-oriented test for high school students who are considering careers in engineering, math, science, or technology and helps determine students' strengths and weaknesses.

SOUTHEASTERN CONSORTIUM FOR MINORITIES

For nearly 20 years, the Southeastern Consortium for Minorities in Engineering (SECME) has helped increase the number of minorities studying and earning degrees in engineering, mathematics, and science. SECME students compete for industry and university scholarships.

The organization works directly with high schools and teachers. Its programs help middle-grade teachers meet the needs of minority students and keep those students interested in math and science by doing. SECME's crosscultural workshops show teachers how to understand and value cultural differences in the classroom, making teachers sensitive to students' needs, customs, traditions, and learning styles. Teachers learn how to help students work effectively as a team with classmates from different cultural backgrounds.

SECME's Summer Institute includes an annual student competition among high school finalists from the local and state levels. For information, write to the Southeastern Consortium for Minorities in Engineering (SECME), Georgia Institute of Technology, Atlanta, Georgia 30332-0270.

NATIONAL SOCIETY OF BLACK ENGINEERS

Another organization you will find helpful is the National Society of Black Engineers (NSBE). Through national and regional conferences, NSBE encourages and advises disadvantaged young people to pursue engineering careers. A career fair, resume books, and technical seminars are part of the annual national conference.

Members of this student-run, nonprofit organization visit schools and host junior and high school students on campuses. NSBE presents scholarships to high school seniors and, with support from the corporate sector, NSBE presents scholarships based on scholastic achievement.

For information, write the National Society of Black Engineers (NSBE), 344 Commerce Street, Alexandria, Virginia 22314.

ADVANTAGES OF ORGANIZATIONS

If you are a minority student, especially if you are studying engineering or engineering-related courses, you will want to take advantage of all the help that is available. Often this includes summer programs before the freshman year of college; developmental year programs, which have special courses; and tutoring. Taking part in programs like this is time well spent; make a special effort to look for help and take advantage of opportunities.

You will want to look for student chapters of national associations on your campus or for senior chapters nearby. Join them and attend meetings. Organizations such as the Society of Manufacturing Engineers (SME), the Institute of Industrial Engineers, and the Society of Women Engineers all have extremely active chapters. In fact, the Society of Women Engineers had roughly 9,000 student members in 1992. Taking an active, leadership role in organizations like these not only gives you valuable experience but also helps you make contacts that can be helpful as you enter the job market.

Special Publications

A new series of ethnic titles, often on reference shelves at public and college libraries, has extremely useful information. Available from Gale Research, Inc., and updated periodically, the series includes:

- *Asian Americans Information Directory*
- *Black Americans Information Directory*

- *Hispanic Americans Information Directory*
- *Native Americans Information Directory*

Typical of the books in the series, *Hispanic Americans Information Directory*, a biennial book published in October of odd years, is a guide to approximately 4,700 organizations, agencies, institutions, programs, and publications concerned with Hispanic American life and culture. Included are listings of associations; awards, honors, and prizes; bilingual and migrant education programs; the top 500 businesses; cultural organizations; government agencies and programs; libraries; newspapers and periodicals; research centers; and videos.

Organizations and publications are described in detail. For instance, the Society of Hispanic Professional Engineers—National Newsletter describes reports on activities, programs, and workshops of the association; announcements of educational and job opportunities and scholarships; affiliate association and member news; calendar of events; and columns.

Other Organizations

Here are several additional organizations that specifically target minorities and offer career information:

Hispanic Society of Engineers and Scientists (HSES)
P.O. Box 1393
Richland, WA 99352

Mexican-American Engineering Society (MAES)
P.O. Box 3520
Fullerton, CA 92634

Mid-America Consortium for Scientific and Engineering Achievement
c/o Engineering Department
Durland Hall, Kansas State University
Manhattan, KS 66506

Mid-American Chinese Science and Technology Association
    (MACSTA)
    P.O. Box 4528
    Naperville, IL 60567

National Action Council for Minorities in Engineering (NACME)
    3 West 35th Street
    New York, NY 10001

National Council of La Raza (NCLR)
    955 L'Enfant Plaza SW, Suite 4000
    Washington, DC 20024

Society for the Advancement of Chicanos and Native Americans in
    Science (SACNAS)
    P.O. Box 30030
    Bethesda, MD 20814

Society of Spanish Engineers, Planners and Architects
    P.O. Box 75
    Church Street Station
    New York, NY 10007

ON THE JOB

If you have successfully completed your schooling, especially
if you have an engineering degree, finding a job in the robotics
industry and advancing your career may still not be easy. However, your chances, if you are a minority applicant, are as good
as anyone else's. Increasingly, more and more firms are coming
to accept the reality of cultural diversity.

The Bureau of Labor Statistics predicts a more culturally diverse work force in the next decade, with increased participation
from women and minorities. For instance, BLS believes that
within the work force, by the year 2000, the group with the largest
numerical growth will be women in the prime working years, ages
25 to 54. This group is projected to increase by ten million,

compared with the seven million increase in prime age men. In fact, BLS says, "prime age women not only would account for the largest labor force increase, but would also have the highest rate of growth."

Blacks in the labor force are expected to make up 12 percent of workers, up one percentage point from 1988—reflecting a higher growth rate than BLS projects for the overall labor force. The numbers of Asians are growing at a rate of 3.6 percent annually— higher than either the black or white rate of increase but below the rate increase projected for Hispanics. Between 1990 and 2000, says BLS, there will be an increase of 5.3 million Hispanics in the labor force—to 14.3 million in 2000, and Hispanics are projected to constitute 10 percent of the labor force, up three percentage points from 1988.

What do these figures mean for opportunities in robotics careers? More importantly, what do they mean for you? No one can promise you a job or guarantee you a successful career. However, it certainly seems that if you have the desire to succeed in robotics, related technology, or similar fields of electronics and engineering—and the qualifications to perform effectively— you have an excellent chance for success, regardless of race or ethnic background.

CHAPTER 11

INTERNATIONAL OPPORTUNITIES

No book on robotics would be complete without a discussion of Japan's strong position in robotics and related technologies. In fact, Japan represents a strong magnitude and force in the U.S. robot market.

According to the International Federation of Robotics, an organization that compiles information supplied by national robotic societies, at the end of 1991, Japan had approximately 324,895 industrial robots. The former Soviet Union had an estimated 65,000 industrial robots, and the United States had approximately 44,000 industrial robots.

Other countries with sizable numbers of industrial robots installed at the end of 1991 included Germany (34,140), Italy (14,700), France (9,808), and the Czech and Slovak Federal Republic (7,211).

More information on robot data can be obtained from the International Federation of Robotics (IFR), Box 5506, S-114 85, Stockholm, Sweden. All IFR robot statistics from 1980 are available from IFR in a data base produced in Lotus 1-2-3, version 3. Data is broken down by countries, application areas, industrial branches, and types of robots. The IFR also publishes a *Yearbook of Industrial Robot Statistics*.

As you look at these figures, you must keep in mind several key points: A *robot* is not necessarily a robot—that is, all countries do not define *robot* in the same way. The difference in definitions can distort statistics that compare robot populations in various countries.

For instance, the Robotic Industries Association says, "A robot is a reprogrammable, multifunctional manipulator designed to move material, parts, tools or specialized devices through variable programmed motions for the performance of a variety of tasks." It uses this definition to estimate the number of industrial robots in the United States. Japan, however, includes mechanically fixed manipulators in its definition of *robot*. The motions of these manipulators are set by mechanical cams installed in the factory. The manipulators are typical of "hard automation," in contrast to the programmable industrial robots, often thought of as "flexible automation."

Since the Japanese definition is different, any comparisons should be made cautiously. How can the value of industrial robots be calculated? Are robots counted by themselves, or are other devices that are needed to install the robot included in the count? Sometimes a manufacturer produces only the robot. At other times, the manufacturer produces an entire system. In determining value (and numbers of robots), is a simple robot that operates in three axes counted the same way as a much more complicated, six-axes, intelligent robot?

ROBOTS IN JAPAN

An estimated 50,000 robots are installed in Japan each year, according to Donald A. Vincent, executive vice-president of Robotic Industries Association. "It's no accident that Japan is such a strong competitor in so many manufacturing industries," he says. "Their companies are willing to make a long-term commitment to robotics and advanced automation."

Robot Applications in Japan

One of the largest robot markets has been the automotive industry. Companies like Toyota and Nissan are using robots to install rear windows, spare wheels, and clusters of rear lamps. They are inserting body struts and tightening screws. Robots mount wheels on the body and measure wheel alignment. In addition, Toyota has used robots to assemble components such as hydraulic power-steering pumps and control valves.

Another major Japanese user has been the electrical industry. Because robots are used extensively, Japanese manufacturing of video and audio tape players and recorders has grown rapidly.

Products such as electric fans, dot-matrix and laser printers, and electric shavers are being manufactured with robot assembly lines. Robots are also used extensively to load and assemble electronic components on printed circuit boards.

Japanese assembly robots are far simpler in design than their U.S. counterparts. Compared to an American robot, the Japanese robot has fewer joints. Consequently, it can move faster and more precisely.

In addition—and this is a key factor—many Japanese products have been redesigned to make them easier to assemble. Because Japanese robots are simpler, some industry analysts believe it is easier and quicker to reprogram them. Their high speed and accurate output have helped Japanese manufacturers reduce inventory, thus cutting production costs.

Japan Industrial Robot Association

Partially subsidized by the Japanese government, the Japanese Industrial Robot Association (JIRA) is a coordinating association that helps the robot industry. It provides research on technology and markets and sponsors exhibitions. JIRA's international robot technology center works with the Robotic Society of Japan—a group of manufacturers, students, and others who are interested in robotics, as well as with the Japan Machine Tool Builders Association.

ROBOTS IN CANADA

The Canadian government is interested in manufacturing technology, especially because of the opportunities it offers for growth in trade. A strong automotive parts industry in Canada is centered in Ontario in such areas as Kitchener-Waterloo, Toronto and its suburbs, Windsor, and Oakville. There are also centers in Quebec, British Columbia, and Nova Scotia.

No one knows exactly how many robots there are in Canada, although Ian Barrie, a Canadian robotics expert, estimated that in 1986, the country had about 1,200 robots in use. Yet two of the government-funded Ontario Centres for Advanced Manufacturing—the Ontario Robotics Centre in Peterborough and the Ontario CAD/CAM Center in Cambridge, both established in 1982—were closed down.

Since Canada has not joined the International Federation of Robotics, it is difficult to know the size of its robot population and other statistics about job opportunities. You can write to the Ministry of Technology and Trade in Ottawa for information; that office can suggest sources which may be able to help.

ASSOCIATIONS

There are a number of trade associations involved with varying aspects of robotics, factory automation, and engineering. Some of them have student membership rates. Most associations sponsor conferences or meetings that students can attend for reduced fees. Nearly all of them produce journals, magazines, newsletters, or other publications. Some offer videotapes that can be borrowed or rented; others have career guidance information.

Several of the major associations are listed below. You will find other organizations to contact listed in appropriate chapters throughout this book, including those mentioned in chapter 10.

Of special importance to you as you explore the idea of working in robotics is membership in organizations that have student chapters. Belonging to a student chapter not only brings you in contact with new, exciting ideas but also gives you a chance to take a leadership role. If you are a part of a national or international organization, you will make contacts that can help you later on. In addition, you will keep up-to-date on new technology developments and be aware of what is happening throughout the industry, especially as it relates to job prospects.

ROBOTIC INDUSTRIES ASSOCIATION (RIA)

Founded in 1974, the Robotic Industries Association is the only trade association in North America organized specifically to serve the field of robotics. RIA is dedicated to the exchange of technical and trade-related information between robot manufacturers, distributors, corporate users, accessory equipment and systems suppliers, consultants, research groups, and international organizations. RIA is the common ground where these groups can come together to discuss problems and solutions dealing with the implementation of robotic technology.

RIA sponsors the International Robots and Vision Automation Show and Conference—the only North American trade show devoted exclusively to robotics, machine vision, and related automation technologies. RIA also sponsors Machine Vision and Imaging Financial Forum, an event that brings together industry executives with leading financial analysts and venture capitalists to discuss the outlook for the industry.

In addition, RIA sponsors the National Robot Safety Conference and the International Service Robot Congress—devoted exclusively to applications of robots in health care, education, security, food service, household use, and related service tasks.

While membership in RIA is open to companies, individuals who would like to become part of the association join the Global Automation Information Network (GAIN).

AUTOMATED IMAGING ASSOCIATION

This association, part of RIA, promotes the acceptance and productive use of image processing, image analysis, and machine vision technologies. Members are manufacturers of related and peripheral products, integrators, end users, consultants, and research groups directly involved in these technologies.

NATIONAL SERVICE ROBOT ASSOCIATION

Individuals and corporate members who belong to this organization, part of RIA, include researchers, manufacturers, developers, users, hobbyists, and students. The association is concerned with the application of robot technology to human services: health care, education, security, space and undersea exploration, and related nonmanufacturing areas.

For more information, write Robotic Industries Association, 900 Victors Way, P.O. Box 3724, Ann Arbor, Michigan 48106.

SOCIETY OF MANUFACTURING ENGINEERS (SME)

One of the most important associations in the field is the Society of Manufacturing Engineers (SME). Founded in 1952, SME has 75,000 members in 72 countries, and sponsors over 350 senior chapters and 200 student chapters worldwide.

SME annually sponsors more than 150 special programs, 50 technical conferences, 40 expositions, and 200 symposiums and workshops. It is the umbrella organization for seven related associations: Robotics International (RI/SME), Computer and Automated Systems Association (CASA/SME), Association for Finishing Processes (AFP/SME), Machine Vision Association (MVA/SME), Association for Electronics Manufacturing (EM/SME), Composites Manufacturing Association (CMA/SME), and Machining Technology Association (MTA/SME).

SME publishes technical papers on various subjects that can be purchased individually or (often) in various collections. An on-line electronic database search, available through the society's library service, allows members and nonmembers quick access to papers on desired topics.

Within Robotics International (RI/SME), various areas of interest hold conferences and publish papers. Included in robotics are welding, coatings, machine loading and unloading, assembly, casting and foundries, research and development, aerospace, hazardous handling, human factors, vision, education and training, electronics, and military systems.

Machine vision areas of interest include assembly, image processing, inspection, lighting and optics, measurement/metrology, object recognition, optical measurement, process control, quality control, research and development, robotics, and sensors.

SME members who want to join more than one association get a special dual rate. Ask for details.

SME also offers certification through examination as peer group recognition of education and experience. Details are available directly from SME.

For membership rates and more information on SME, write the Society of Manufacturing Engineers, One SME Drive, Dearborn, Michigan 48121.

INSTITUTE OF INDUSTRIAL ENGINEERS (IIE)

Founded in 1948, IIE is an international, nonprofit, professional society with over 35,000 members in the United States and 89 other countries. Its members work in the manufacturing and service industries, government, or as consultants and faculty members. Industrial engineers are actively engaged in such fields as energy conservation, plant design and engineering, systems engineering, production and quality control, performance and operational standards, material flow systems, and operations research. Virtually all of them are involved in productivity and quality improvement through systems integration.

IIE is organized into two societies, eleven divisions, and ten interest groups. The institute has 200 senior chapters in the United States and around the world and 135 university chapters. Publications include *Industrial Engineering* and *Industrial Management* magazines, journals, books on current topics, and conference proceedings.

For more information, contact the Institute of Industrial Engineers, 25 Technology Park/Atlanta, Norcross, Georgia 30092.

SPIE—THE INTERNATIONAL SOCIETY FOR OPTICAL ENGINEERING

This society is dedicated to advancing engineering and scientific applications of optical, electro-optical, and optoelectronic instrumentation, systems, and technologies. It has established a Technical Working Group on Robotics and Machine Perception.

The group emphasizes robotic sensing and its broader relationships to machine perception, planning, and control systems. Research and development topics of interest include computer vision, sensor fusion, machine metrology and calibration, task planning, intelligent control, real-time sensing and control architectures, 3-D modeling and simulation, man-machine interface, robotic navigation, and telerobotics. Engineering issues include software methodology, task benchmarks, and design standards. Applications include assembly, inspection, servicing, surveillance, science exploration, and sample collection.

The group publishes *Robotics,* a quarterly newsletter with articles that cover industry news, techniques, tools, and standards issues. The Robotics and Machine Perception Group holds its annual meeting in conjunction with SPIE's International Symposium on Advances in Intelligent Robotic Systems.

For more information on SPIE and the working group, contact SPIE—The International Society for Optical Engineering, P.O. Box 10, Bellingham, Washington 98227-0010.

PERIODICALS

If you want to work in robotics or in related technologies, it is extremely helpful to start learning all you can about this rapidly changing field. You can benefit greatly from keeping abreast of this complex technology not only in engineering developments but also in broader issues that factory automation raises.

Periodicals in the field of manufacturing engineering, including those magazines and reports that concentrate on robotics-related news, continue to discuss such topics as justifying costs, retraining workers in skills needed for new technology in factory automation, retrofitting existing plants, and implementing safety procedures. Of course, new products and new applications of existing products are covered. Periodicals also report on the economics of the robot industry as well as on how major companies are having an impact on jobs and products.

Keeping up with industry developments by reading several periodicals regularly is a good idea. You will be able to assess what is happening and how these developments can directly affect you. For instance, there was early enthusiasm about robotics and industry growth during the early 1980s. However, most experts now believe that robotics should be considered in the broader context of industry's move to computer-aided manufacturing. You may want to consider, then, schools that stress broader electronics and computer training rather than limit yourself purely to robotic work.

HOW TO KEEP UP

There are several ways to keep up with the literature. One good system is to check your public or university library for *Applied Science and Technology Index* and the *Business Periodicals Index*. These specialized indexes are published by the same company that issues the familiar *Readers' Guide to Periodical Literature.* They are updated monthly, with three months' worth of citations grouped together in quarterly paperback volumes. Eventually, all the updates are combined in an annual volume. Both indexes are cross referenced, so that if you look for the topic "Robotics," you will be referred to articles on related topics.

In *Applied Science and Technology Index,* for example, when you look up the robot industry, you will be told to "see also" the headings "robotic painting" and "robotic welding." When you look up "robots," you will be referred to articles on industrial robots, inspection robots, and mobile robots. Under "robot programming," you will be referred to articles on assembly robots—programming; industrial robots—programming; manipulators—programming; underwater robots—programming; and welding robots—programming. If it seems as if some of these topics overlap, they do. You may find the same citation under several different headings.

For ease in using the indexes, you may want to photocopy the pages listing the citations. Then you can take them home and review them, deciding on what you really want most to read. You can also ask your reference librarian for a list of what publications the particular library you are using subscribes to and whether nearby libraries have some of the others. A comparison of the lists will show you quickly whether the articles which interest you are easily obtainable.

It is highly unlikely that the average public library will subscribe to all these specialized journals and trade publications. However, many libraries belong to services which can obtain copies of individual articles, often at no charge or for an extremely inexpensive fee. There are copyright rules which the library may use as guidelines to limit the number of articles from any particular journal or issue.

In the front of *Applied Science and Technology Index* and *Business Periodicals Index,* you will find a list of the periodicals covered by that particular index, along with the address of the magazine. Costs of a subscription are listed in *The Gale Directory of Publications and Broadcast Media,* which is updated yearly and available at the reference desk of your library. Costs usually are listed for a sample copy so that you can write to the publisher directly and enclose your check or money order.

If you are asking a publisher outside your own country to send you a sample copy, talk with your post office or postal authority about buying international money orders and international postal reply coupons. Publishers abroad may be reluctant to accept personal checks. Also, sending them the international postal reply coupons (which you buy at your local post office) lets them exchange the coupons for postage in their own country's stamps. Don't forget to let the publisher know whether you want the sample sent by surface mail or by air. If it is to come by air, be sure to enclose enough international postal reply coupons to cover the cost of airmail postage, and say so explicitly in your letter.

The following periodicals will be especially helpful to you as you expand your awareness of the robotics industry and related technology.

Automotive Engineering
  c/o Society of Automotive Engineers
  400 Commonwealth Drive
  Warrendale, PA 15096

Articles, photos, and drawings about advances in technology in the auto industry. The magazine monitors government action, including developments in technology that strengthen U.S. industry's competitiveness.

Communications of the ACM
  Association for Computing Machinery
  1515 Broadway
  New York, NY 10036

This monthly publication focuses on computing science for the practitioner. Articles cover all disciplines, including artificial intelligence (AI), language processing, multimedia, and cryptography. Columns feature information on legal issues, social issues, education, public policy, and internal perspectives in computing.

Computer Design
  PennWell Publishing Company
  Advanced Technology Group
  One Technology Park Drive
  P.O. Box 990
  Westford, MA 01886

A technical publication that concentrates on system design and integration.

Computer
  c/o The Institute of Electrical and Electronics Engineers
  IEEE Computer Society
  10662 Los Vaqueros Circle
  Los Alamitos, CA 90720

Highly technical, with an occasional special issue or article on robotics and related technologies.

Control Engineering
Technical Publishing
875 Third Avenue
New York, NY 10022

Designfax
Huebcore Communications, Inc.
1355 Mendota Heights Road
Suite 210
Mendota Heights, MN 55120

Professional design engineers stay current with this useful, practical, easy-to-read source on changing technologies and new products. You will find information on product design trends and new design technology in electrical and electronic design, fluid power, materials, mechanical components, CAD, computer, and other engineering-related devices and equipment.

Similar publications from the same company include *Metlfax* (metal cutting, metal forming, fabrication, quality, and automation) and *Medical Equipment Designer*. Only qualified professionals receive these controlled circulation publications, but sample copies may be purchased.

Design News
Cahners Publishing Company
8773 South Ridgeline Boulevard
Highlands Ranch, CO 80126

This magazine blends articles with primarily technical vocabulary with easy-to-read illustrated features on industrial topics. Its audience is primarily design engineers. New product information and technology business briefs are included for a variety of industries.

Electronic Design
611 Route 46 West
Hasbrouck Heights, NJ 07604

A worldwide audience of engineers and engineering managers uses this fairly technical magazine to keep up with developments in processor chips, power op amps, flash converters, and other such subjects.

Electronics
　　1100 Superior Avenue
　　Cleveland, OH 44114-2543

A worldwide technology biweekly that looks similar to *Time* or *Newsweek* and covers many topics in brief, including CAD/CAM, semiconductor technology, and computer technology.

Engineering Digest
　　Canadian Engineering Publications Limited
　　5080 Timerlea Boulevard
　　Suite 8
　　Mississauga,Ontario, Canada L4W 5C1

Canadian and U.K. news of interest to engineers is highlighted, along with new product descriptions. Each main feature article is summarized in English and in French.

IEEEExpert
　　10662 Los Vaqueros Circle
　　P.O. Box 3014
　　Los Alamitos, CA 90720-1264

This highly technical publication covers intelligent systems and their applications. It is published by the IEEE Computer Society and covers technologies relevant to robotics.

Industrial Engineering
　　Institute of Industrial Engineers
　　25 Technology Park/Atlanta
　　Norcross, GA 30092

This publication is concerned with the design, improvement, and installation of integrated systems of people, material, information, equipment, and energy. Many articles on automating production lines are included in colorful, easy-to-read form.

Industrial Finishing
Hitchcock Publishing Company
191 South Gary Avenue
Carol Stream, IL 60188

Deals primarily with the paint and coatings industry, including manufacturing and application.

Journal of Metals
Metallurgical Society of AIME
420 Commonwealth Drive
Warrendale, PA 15086

Explores materials science and engineering with articles of interest to those studying or working in the field. Annual review of extractive metallurgy.

Laser Focus World
Pennwell Publishing Company
One Technology Park Drive
P.O. Box 989
Westford, MA 01886
Circulation Department
P.O. Box 3004
Tulsa, OK 74101

Global electro-optic technology and markets, including lasers, optics, detectors, instrumentation, and fiber-optics.

Machine Design
Penton Publishing
1100 Superior Avenue
Cleveland, OH 44114-2543

Drawings, color photos, and graphs help explain complicated concepts of interest to those designing industrial machinery.

Managing Automation
Thomas Publishing Company
Five Penn Plaza
New York, NY 10001

This publication considers itself the magazine of computer integrated manufacturing. Current news of industry developments and companies, an international report on automation, a calendar of events, and a regular section of industry opinion from leaders in manufacturing automation integrate technical and business information.

Manufacturing Engineering
Society of Manufacturing Engineers
One SME Drive
Dearborn, MI 48121

Materials Evaluation
American Society for Nondestructive Testing
1711 Arlingate Plaza
P.O. Box 28518
Columbus, OH 43228-0518

This publication is the official journal of the American Society for Nondestructive Testing. It includes information on upcoming meetings of various related associations, new patents, and methods of nondestrutive testing across a variety of industries.

Mechanical Engineering Magazine
American Society of Mechanical Engineers
345 East 47th Street
New York, NY 10017

Modern Machine Shop
6600 Clough Pike
Cincinnati, OH 45244-4090

Robot systems are reviewed occasionally in the magazine's "Modern Equipment Review" section.

Modern Materials Handling
  Cahners Publishing Company
  221 Columbus Avenue
  Boston, MA 02116

Covering topics in management, systems, and equipment, this magazine includes features about robots in manufacturing and warehousing applications.

Plant Engineering
  Cahners Publishing Company
  1350 East Touhy Avenue
  Des Plaines, IL 60018

This magazine is directed to people in the manufacturing industry who perform the plant engineering function. Such people are responsible for the maintenance, repair, modification, and operation of plant facilities, equipment, and systems. There are occasional articles on robotics.

Robotics
  SPIE–The International Society for Optical Engineering
  P.O. Box 10
  Bellingham, WA 98227-0010

This newsletter is published by SPIE's Technical Working Group on Robotics and Machine Perception. Articles discuss the many roles sensors and signal processing play in robotics design and applications.

Robotics Today
  Society of Manufacturing Engineers
  One SME Drive
  Dearborn, MI 48121

Quarterly publication for members of the Robotics International (RI) division of the Society of Manufacturing Engineers.

Robotics World
  6151 Powers Ferry Road, NW
  Atlanta, GA 30339-2941

This controlled-circulation trade publication goes only to qualified recipients (i.e., people who are currently working in the robotics field and whose job titles match the target audience).

Each issue includes product application articles. An annual *Robotics World* directory lists robot manufacturers, robotic/vision systems integrators, suppliers' applications, robotic research, consultants, and training programs. The magazine is available in libraries that maintain a paid subscription.

Sensors
  Helmers Publishing, Inc.
  174 Concord Street, P.O.B. 874
  Peterborough, NH 03458-0874

The journal of machine perception, *Sensors* is edited for design, production, and manufacturing engineers involved in the detection, control, and measurement of specific physical properties and conditions. Editorial content focuses on using sensing devices to increase efficiency, economy, and productivity in applications that range from manufacturing systems to process control and from aircraft to consumer products. The magazine emphasizes new developments in sensor technology and on innovative applications of existing sensing methods.

Tappi Journal
  Technical Association of the Pulp and Paper Industry, Inc.
  Technology Park/Atlanta
  P.O. Box 105113
  Atlanta, GA 30348-5113

This journal includes information on process control and monitoring in the pulp and paper and related industries (corrugated containers, packaging, nonwovens).

Tooling & Production
    Huebcore Communications Inc.
    29100 Aurora Road
    Suite 200
    Cleveland, OH 44139

The magazine of manufacturing technology and management, *Tooling and Production* provides in-depth coverage of advances in manufacturing automation and control. Recent articles include "The Promise of Robotics," a look at the future of robots in manufacturing.

Vision Newsletter
    Machine Vision Association of SME
    Society of Manufacturing Engineers
    One SME Drive
    P.O. Box 930
    Dearborn, MI 48121-9939

This quarterly newsletter includes information on developments in the machine vision industry and in vision applications and research.

Welding Design & Fabrication
    Penton/IPC Inc.
    P.O. Box 95759
    Cleveland, OH 44101

Industry news, quality control, research, feature articles about advances in welding, and case histories are highlighted in this magazine of welding management and technology. Of special interest: salary, age, education level, and experience distribution of welding engineers are all included in the magazine's semiannual survey.

Welding Journal
  American Welding Society
  P.O. Box 351040
  Miami, FL 33135

Published monthly by the American Welding Society, the *Welding Journal* offers feature articles, high-tech research, and practical "how-to" articles about all aspects of welding, including automation and robotics. Opportunities for young people entering the fields of welding and welding engineering are covered regularly in the magazine.

BOOKS

The following books on robotics and related technologies are recommended. Check *Subject Guide to Books in Print* for additional titles.

Fifth International Service Robot Congress: Proceedings, Ann Arbor, Michigan, Robotic Industries Association, 1988.

1992 IEEE International Conference on Robotics and Automation (3 vols.), Los Alamitos, California, IEEE Computer Society Press, 1992.

International Robot and Vision Automation Conference Proceedings, Ann Arbor, Michigan, Robotic Industries Association, 1991.

Robotics, rev. ed. (Understanding Computers Series), Alexandria, Virginia, Time-Life, 1991.

MVA/SME Machine Vision Industry Directory, Dearborn, Michigan, SME, 1991.

RI/SME Robotics Research and Development Lab Directory, Dearborn, Michigan, SME, 1991.

Vision '90 Conference Proceedings, held Nov. 12–15, 1990, Detroit, Michigan, Dearborn, Michigan, Society of Manufacturing Engineers, 1990.

Asfahl, C. Ray, *Robots and Manufacturing Automation,* New York, John Wiley & Sons, Inc., 1992.

Bolhouse, Valerie, *Fundamentals of Machine Vision,* Dearborn, Michigan, self-published, 1991.

Desrochers, Alan A., ed., *Intelligent Robotic Systems for Space Exploration,* Hingham, Massachusetts, Kluwer Academic Publishers, 1992.

Dorf, Richard, and Shimon Y. Nof, eds., *Concise International Encyclopedia of Robotics,* New York, John Wiley & Sons, Inc., 1990.

Galbiati, L., Jr., *Machine Vision and Digital Image Processing Fundamentals,* New York, Prentice Hall, 1990.

Goetsch, D., *Advanced Manufacturing Technology Book,* Dearborn, Michigan, SME and Delmar, 1990.

Hall, E. L., et al., eds., *Expert Robots for Industrial Use,* vol. 1008, Boston, SPIE-Society of Photo-Optical Instrumentation Engineers, 1989.

Henderson, T.C., *Traditional and Non-Traditional Robotic Sensors,* New York, Springer-Verlag, 1990.

Hoshizaki, Jon, and Emily Bopp, *Robot Applications Design Manual,* New York, John Wiley & Sons, Inc., 1990.

Hunt, V. Daniel, *Understanding Robotics,* San Diego, California, Academic Press, 1990.

Iyengar, S.S. and Alberto Elfes, *Autonomous Mobile Robots: Perception, Mapping, and Navigation—vol. 1,* Los Alamitos, California, IEEE Computer Society Press, 1991.

———, *Autonomous Mobile Robots: Control, Planning, and Architecture —vol. 2,* Los Alamitos, California, IEEE Computer Society Press, 1991.

Jarvis, R.A., ed., *Robots: Coming of Age,* New York, Springer-Verlag, 1989.

Khatib, Oussama, John J. Craig, and Tomas Lozano-Perez, eds. *The Robotics Review 2,* Cambridge, Massachusetts, MIT Press, 1992.

Leatham-Jones, B., *Elements of Industrial Robotics,* East Brunswick, New Jersey, Nichols Publishing Co., 1989.

McKerrow, Philip, *Introduction to Robotics,* Redding, Massachusetts, Addison-Wesley, 1991.

National Center for Manufacturing Sciences, *Competing in World-Class Manufacturing: America's 21st Century Challenge,* Homewood, Illinois, Dow-Jones Irwin, 1990.

Pingry, J., *Practical Machine Vision,* Arlington, Massachusetts, Cutter Information Corporation, 1987.

Reichgelt, Hans, *Knowledge Representation: An AI Perspective,* Norwood, New Jersey, Ablex Publishing, 1991.

Salant, Michael A., *Introduction to Robotics,* New York, McGraw-Hill, Inc., 1988.

Sharon, D., J. Harstein, and G. Yantian, *Robotics and Automated Manufacturing,* East Brunswick, New Jersey, Nichols Publishing Co., 1989.

West, Perry, *Machine Vision Lighting and Optics,* Campbell, California, Automated Vision Systems, Inc., 1991.

Zuech, N., *Applying Machine Vision,* New York, John Wiley & Sons, 1988.

Zuech, N., ed., *Gaging with Vision Systems,* Dearborn, Michigan, Society of Manufacturing Engineers, 1987.

———, *Machine Vision: Capabilities for Industry,* Dearborn, Michigan, Society of Manufacturing Engineers, 1986.

VGM CAREER BOOKS

VGM Career Horizons
a division of NTC *Publishing Group*
4255 West Touhy Avenue
Lincolnwood, Illinois 60646-1975